D1216819

THE QUALITATIVE-QUANTITATIVE DISTINCTION
IN THE SOCIAL SCIENCES

BOSTON STUDIES IN THE PHILOSOPHY OF SCIENCE

Editor

ROBERT S. COHEN, *Boston University*

Editorial Advisory Board

ADOLF GRÜNBAUM, *University of Pittsburgh*

SYLVAN S. SCHWEBER, *Brandeis University*

JOHN J. STACHEL, *Boston University*

MARX W. WARTOFSKY, *Baruch College of the City University
of New York*

VOLUME 112

THE QUALITATIVE-QUANTITATIVE DISTINCTION IN THE SOCIAL SCIENCES

Edited by

BARRY GLASSNER

Department of Sociology, University of Connecticut

and

JONATHAN D. MORENO

Division of Humanities in Medicine, SUNY Health Science Center at Brooklyn

KLUWER ACADEMIC PUBLISHERS

DORDRECHT / BOSTON / LONDON

Library of Congress Cataloging-in-Publication Data   CIP

The Qualitative–quantitative distinction in the social sciences /
  edited by Barry Glassner and Jonathan D. Moreno.
    p. cm. – (Boston studies in the philosophy of science : v.
112)
    ISBN 90–277–2829–1
    1. Social sciences–Methodology.   I. Glassner, Barry.
II. Moreno, Jonathan D.   III. Series.
Q174.B67 vol. 112
[H61]
501 s–dc 19
[300'.72]   88–23159
                                                CIP

ISBN 90–277–2829–1

Published by Kluwer Academic Publishers,
P.O. Box 17, 3300 AA Dordrecht, The Netherlands

Kluwer Academic Publishers incorporates
the publishing programmes of
D. Reidel, Martinus Nijhoff, Dr W. Junk and MTP Press.

Sold and distributed in the U.S.A. and Canada
by Kluwer Academic Publishers,
101 Philip Drive, Norwell, MA 02061, U.S.A.

In all other countries, sold and distributed
by Kluwer Academic Publishers Group,
P.O. Box 322, 3300 AH Dordrecht, The Netherlands.

H
61
.Q34
1989

printed on acid free paper

All Rights Reserved
© 1989 by Kluwer Academic Publishers
No part of the material protected by this copyright notice may be reproduced or
utilized in any form or by any means, electronic or mechanical,
including photocopying, recording or by any information storage and
retrieval system, without written permission from the copyright owners.

Printed in the Netherlands

WIDENER UNIVERSITY
WOLFGRAM
LIBRARY
CHESTER, PA.
DISCARDED
WIDENER UNIVERSITY

Dedicated to the memory of
Richard S. RUDNER (1921–1979)
Professor of Philosophy, Washington University
Editor, Philosophy of Science (1959–1974)
Author, Philosophy of Social Science (1966)

TABLE OF CONTENTS

BARRY GLASSNER AND JONATHAN D. MORENO

INTRODUCTION:
QUANTIFICATION AND ENLIGHTENMENT

1. QUANTIFICATION AND SOCIAL CHANGE

Contemporary students of the social sciences are well-acquainted with two claims about the role of quantitative techniques in those fields. One is that quantification is essential for an objective and rigorous investigation of the social no less than the 'natural' domain; another is that no description of a social world or an aspect of one can be complete without some qualitative appreciation of relevant properties of the territory. Our purpose here is not to rehearse the several arguments and accounts that could be given in support of or in opposition to one or the other of these not incompatible claims, but rather to show how their familiarity tends to conceal a vast array of presuppositions that can be felicitously displayed through an historical and philosophical analysis of their content.

One observation about the above propositions that immediately presents itself concerns a feature they have in common. Each suggests a somewhat detached relation to any social situation that is its object. That this has not always been the case is easily recalled: eighteenth-century social scientists (the expression is used in full awareness of its anachronistic tendency), virtually identified their progress with that of society, arguing an interactive influence of social and social scientific improvement. Further, the measure of this progress was the increased domination of quantification in the practice of social science, such that the values of social change were determinably numerical and greater facility with quantitative technique augured social improvement.

In a word, quantification, once an augury of change, became in the Enlightenment a *creator* of change. In one sense, then, the history of the social sciences has not always associated quantification with 'value freedom,' though in another sense of the term, Enlightenment social scientists were even more convinced of the value freedom of their numerical results than the moderns. That is, Enlightenment social science was not value free in the sense that it disassociated itself from prescriptions about progressive social policies; quite the contrary. However, the certainty these thinkers had as to

1

Barry Glassner and Jonathan D. Moreno (eds.),
The Qualitative-Quantitative Distinction in the Social Sciences. 1–12.
© 1989 *by Kluwer Academic Publishers.*

the power of their results with respect to making social progress feasible relied on a classical ancient notion of the objective merit of quantification, another sense of value freedom. In our own time we have come somewhat indistinctly to wonder about the objectivity of quantification, not to mention the meaning and value of objectivity itself.

The eighteenth century will turn out to be a pivot upon which the rest of the story of quantification in social science can be said to turn. It will be important to see what the much-vaunted objectivity of quantification has meant over the years, how its underlying rationale has remained remarkably stable while its implications shifted, and how this rationale has helped to foster the confusion from which we still suffer concerning the relationship between qualitative and quantitative methods in social science. The point is not an indictment of any particular social science methodology, but rather a critical examination of certain conceptions of the significance of any methodology as they appear in these developments.

2. QUANTIFICATION AND OBJECTIVITY

Plato articulated a powerful intuition we seem to have about the preeminence of measurement. At one point in the *Republic* Socrates considers the conditions that can provoke reflection, as 'when perception yields a contradictory impression, presenting two opposite qualities with equal clearness...' Sight cannot satisfactorily distinguish between the sizes of one finger and another, for example, for 'sight perceives both big and little; only not as separate, but in a confused impression.' In response, intelligence invokes 'the help of reason with its power of calculation...' Socrates goes on to note that 'number is the subject of the whole art of calculation and of the science of number' and that 'the properties of number appear to have the power of leading us towards reality' precisely because they are more reliable than the properties of mere perception (Plato, 239–241).

Plato's familiar ruminations on this matter follow his adumbration in the cave allegory of the lesson of the analogy of the line. Access to mathematical forms is intellectual, superior to perceptual activity, and a necessary preparation for knowledge of the forms of things themselves. Hence mathematical calculation helps us come closer to awareness of the objects of intellect, the only true objects there are. The overall impression is that, short of insight to the forms themselves, mathematical knowledge is the closest we can come to knowing the true nature of things, while 'knowledge' of qualities is superficial and finally vain.

Without of course adopting a Platonic metaphysics, the eighteenth-century *philosophes* were Grecophiles who regarded the Athenian philosophers as their intellectual forbearers and mentors. So powerful was their identification with classification that ancient ideas were taken as keys to the design of the modern world, but usually the ideas were taken separately and as divided from their systematic context. The power of number was an idea the Enlightenment thinkers deployed with their legendary passion and vigor, particularly as an instrument for social reconstruction. It is no exaggeration to say that the role of quantities in contemporary social scientific theorizing cannot be understood with any depth absent a recollection of the philosophes' axial development of the notion of quantification.

It is a commonplace that for the philosophes progress required releasing human abilities to have power over nature. A prerequisite for this power was knowledge of the underlying causes of natural events, knowledge that required quantitative precision. Enlightenment thinkers were sufficiently aware of themselves as products of their time to appreciate the importance of a liberal social environment to the knowledge enterprise; the supposition that the reverse is also the case, that enhanced knowledge could advance social conditions, came easily. The same conviction justifies the conduct of social science in our own day, with the critical difference that we no longer believe that our theoretical breakthroughs will eventuate in exactly those social innovations that would carry us in the direction which we have determined in advance to be the good society. The philosophes, on the other hand, thought they knew more or less what the contours of the good society were, and were certain that mathematically expressed knowledge was at least a necessary condition of its becoming a reality.

Helvetius' view that the study of virtue in general requires the demonstrative character of geometric calculation was typical. But the philosophers of the eighteenth century were by no means united on the primacy of quantification for all fields of study. In chemistry, for example, there was the famous controversy about material change. On one side were those who preferred an atomic/vacuum theory of the movements of particles as alterations in the time-space coordinates of hard atoms, a notion susceptible to quantification. On the other were those who argued that some basic set of qualitative elements, like Aristotle's hot, moist, cold, and humid, were influenced by a kind of life force or ether, conditions of irregular change which would make mathematization difficult at best. Voltaire regarded the Newtonian mathematical synthesis in physics and chemistry as a victory for Platonists and looked forward to similar events in chemistry, while Diderot's *Encyclopedie* emphasized medieval technologies like blending and mixing as much as measurement (Kiernan, 1st ed., 1968, 42; 2nd ed., 1973, 166–7).

3. QUANTIFICATION AND ENLIGHTENMENT

Among eighteenth-century proponents of quantification in social science Condorcet was the most exuberant; such was his radicalism on so many matters, as on the need to destroy religion to ensure human progress, that even his ally Voltaire flinched (Kiernan, 1973, 113). But it was in Condorcet, one historian has written, that 'all the streams of the Enlightenment united in a single impassioned flow' (Sampson, 119). Condorcet came at the right time to be the philosophe of the French Revolution as well as the first idealogue of the nineteenth century. He carried forward the humanism of Locke and Condillac thus uniting the British and French Enlightenments. His *Esquisse d'un tableau historique des progrès de l'esprit humain* gave the most 'integrated expression' (Kiernan, 1973, 108) to the idea of progress by joining the competing approaches of the physical sciences and the life sciences alluded to above in the debate on chemistry. His vision of continued and unlimited human progress sustained him even as he awaited Robespierre's guillotine.

Condorcet is credited with establishing, and perhaps introducing, the term 'social science' (Baker, 211–226). His conception of this field of study was inextricably linked with his idea of progress, which is the elimination of error by a large number of individuals in society. The judgments that guide conduct are probabilistic, but too often they rely on 'mechanical sentiment,' 'crude and uncertain guesses,' 'absolute truths,' or 'the seductive influences of the imagination, of interest, or of the passions' (Condorcet, 184–194). Without rational calculation social planning is a farce.

…[T]he point would soon be reached where all progress becomes impossible without the application of rigorous methods of calculation and of the science of combinatorial analysis; and the advances of the moral and political sciences – as of the physical sciences – would soon come to a halt (Condorcet, 185).

Condorcet emphasized that the science of 'social mathematics' 'can make progress only to the extent that it is cultivated by mathematicians who have thoroughly studied the social sciences,' for their expertise is required in order to bring the social sciences to their proper level. But ordinary people can also benefit from the 'practical utility' of these methods if they take the time to enlighten themselves. A knowledge of the causes affecting the formation of prices, for instance, can make for a more intelligently managed and stable social order: 'It will be seen how much of this science would contribute, if it were more widespread and better cultivated, both to the happiness and to the perfection of the human race' (Condorcet, 184).

In tracing the historical results of Condorcet's legacy, Keith Michael Baker exposes the irony of August Comte's identification of Condorcet as his 'principle precursor,' for Comte and Saint-Simon transformed progress into an historical rather than a political conception bereft of Condorcet's mathematically organized program for social change (Ch. 6). In this respect contemporary 'positivist' social science is much more in the spirit of Condorcet than Comte; but still more significant is the fact we have completely lost the confidence Enlightenment thinkers had that the introduction of mathematical analysis of social relations would guarantee human progress, even if we could achieve a serviceable definition of the latter term.

4. QUANTIFICATION AND PROGRESS

One familiar critique of the notion that social improvement and mathematical calculation are intertwined has it that they are, in fact, actually antithetical. This was Alexis de Tocqueville's sentiment when he expressed the hope that a sterile rationality would not replace the highly motivated social habits of everyday life, a notion that was paradigmatically expressed in Max Weber's classic attack on the technological rationalization that can destroy the human spirit. Quantification was, by implication, partly responsible for the soulless military applications of science that oppressed the Vietnamese during the sixties, according to counter-culturists. The critique was usually directed at 'technology' and 'objectivity' as for example in Chapter VII of Theodore Roszak's *The Making of a Counter Culture*. Undoubtedly, widespread disappointment in the record of applied mathematical social science, in keeping its promises of social reform, made still more dubious whatever might have remained of the Enlightenment notion in America in the last thirty years.

These observations lead to the second critique of the idea that quantified social analysis tends toward social progress. In the twentieth century there has come to be so little reason to suppose that 'progress' can be significantly defined beyond some limited projects, quite apart from the skepticism whether it has any meaning at all, and a belief that its presence in social science theorizing infects the value-freedom without which the scientist's conclusions are invalid. While individual researchers may harbor the hope and conviction that their work will conduce to social improvement, they are not supposed to select a methodology based upon that consideration, but rather owing to its appropriateness to the area under investigation and the likelihood that it will yield scientifically defensible results. This limitation would have been quite alien to Condorcet, who was far less interested in keeping social science pure than he was in influencing social policy, and who in any case saw no inconsistency between those two aims.

Considering that the origins of contemporary quantitative social science were steeped in Enlightenment values about social progress, there is some irony in the fact that today social scientists who reject the preeminence of quantitative methods, particularly in sociology, are so often thought to be illegitimately advancing their own values for social change, consciously or not. Critics of qualitative methods find them either inherently infected by the unacknowledged preferences of the practitioner or at least more liable to such infection. Often, too, qualitatively oriented social scientists are perceived as social activists who subordinate their objectivity to their convictions. Regardless of how widespread or accurate these views really are, their presence attests to the curious twist that has overtaken the relation between quantification and progress since the Enlightenment.

5. WORDS AND QUALIFICATION

For most practicing qualitative social scientists the stakes are not in any event whether to achieve objectivity or not. They may consider that concept obscure, ideological or passé, but in their research they want to achieve something like what it naively connotes. Rather, it is the *nature* of the objectivity sought which is at issue: whether the better understanding of the social world is in words or numbers.

Thomas Reid suggested in 1748 that until 'our affections and appetites shall themselves be reduced to quantity, and exact measures of their various degrees be assigned, in vain shall we essay to measure virtue and merit by them. This is only to bring changes on words, and to make a show of mathematical reasoning, without advancing one step in real knowledge' (717). On the view of some qualitative social scientists, that day has come, thanks in part to the activities of quantitative social scientists. Ordinary citizens do cognize their lives, at least at times, on one-to-five scales of self-esteem and happiness, their political views in terms of degree of agreement with preformed 'opinions'.

Whether a quantitative culture is best served by a quantitative social science is one question, while qualitative social science is another while qualitative social scientists see themselves more as working in the tradition of Simmel and the two Meads than of Freud, they join him in a march back to the word.

Nothing takes place between them except that they talk to each other. The analyst makes use of no instruments – not even for examining the patient – nor does he prescribe any medicines. If it is at all possible, he even leaves the patient in his environment and in his usual mode of life during the treatment ... And incidentally, do not let us despise the *word*. After all it is a powerful instrument; it is the means by which we convey our feelings to one another, our method of influencing other people. Words can do unspeakable good and cause terrible wounds. No doubt 'the beginning was the deed' and the word came later; in some circumstances it meant an advance in civilization when deeds were softened into words. But originally the word was magic – a magical act; and it has retained much of its ancient power (Freud, 187–188).

There is no lack of words in quantitative social inquiry, much of which depends upon questionnaires, but for the quantitative researcher the words are would-be numbers, while for the qualitative researcher they are the very stuff of knowledge. The quantitative social scientist converts 'male' or 'yes' to '2,' 'slight opposition to abortion' to '1,' and so forth. The qualitative social scientist converts words only into other types of verbal entities. The ethnomethodologist takes a statement about gender or abortion to be an 'account,' the symbolic interactionist inquires into the 'meanings' it has in particular types of transactions, the critical theorist assesses its relation to an 'ideology,' the post-structuralist situates it in a field of 'discourse.'

To put the contrast more sharply, the quantitative social scientist reduces words, the qualitative enlarges them. Questionnaires contain few responses per question and are intended to be as efficiently phrased as possible. They are prerecorded into those categories deemed relevant to the research hypothesis. In contrast, qualitative research etiquette calls for 'probing' (Spradley), the continual request for additional information about a subject's report or an observation in the field. At the data analysis stage as well, words are not eliminated or converted into other symbols, but organized by the addition of analytic marginalia (more words).

6. PHILOSOPHICAL APPROACHES AND FAD'S

There is a lacuna in the Enlightenment view that leads us into some metaphysical assumptions that seem to underlie much of the relevant methodological talk. The advantageous feature of quantification is taken to be its precision, as against the imprecise nature of expressions of quality. But there is evidently a danger that the notion of precision is one that relies in advance upon *mathematical* precision as its paradigm case, while the possibility that there is a kind of precision proper to qualitative discourse is not considered. Or it may be argued that mathematical precision is valued on some further pragmatic grounds, such as the greater reliability of predictive hypotheses couched in quantification, but this again looks like a comment upon the *kind* of reliability, namely mathematical, a prediction is supposed to have.

A different approach which some proponents of strict quantification might be disposed to defend is a sort of unreconstructed Pythagoreanism: scientific description is to be quantitative because the world is 'written' in numbers, for number is that property which anything must possess, come what may. It is important that this position is not defended on pragmatic grounds merely, but rather on more ultimate grounds such as correspondence, perhaps. Of course, qualitative ultimacy could be defended in a similar fashion. Although almost nobody in the social sciences actually takes these positions they have a shadow existence in discussions which seek to ameliorate between purely quantitative and purely qualitative approaches. One text on social science methods insists that 'each methodology complements the other' (Smith, 31), and another that nature is '*both* quality and quantity' (Cantore, 91). Actually these two views are logically unrelated. While avoiding the Scylla of numerical monism or the Charybdis of qualitative monism one is tempted to assume a 'double aspect' doctrine to the effect that each method gives us blind people exactly one-half the elephant. Such methodological liberalism, while admirable in other respects, smacks of what might be called the Fallacy of Ameliorative Dualism (FAD), committed by one wishing to admit to the canon of some field of study two theses without obvious systematic association to one another in an attempt to satisfy advocates on both sides. Instances of FAD's or *ad hoc* pairings are by no means limited to the quality-quantity dispute; they have commonly appeared in the history of philosophy as attempts to resolve metaphysical issues, sometimes as the less satisfactory versions of double aspect theories of the mental-physical dichotomy. Some varieties of 'centrism' in politics commit FAD in trying to swallow dogmas of the Left and Right simultaneously. In the present context, however, the danger is that FADish statements may make us think that by admitting the 'complementarity' of methods some deeper question has been settled. A moment's reflection indicates, for example, that the admission of the complementarity of qualitative and quantitative methods does not imply a similar admission to the effect that qualities and quantities are on an ontological par. The former position, often adopted as a hedge against vexing arguments about methodological ideology in a social science, is irrelevant to the latter underlying metaphysical issue that may be the actual agenda of the disputants within the field. On the other hand, it goes without saying that even if one argued that qualities and quantities were ontologically irreducible categories, methodological complementarity would require further argument.

Underlying the sentiments that give rise to the fallacy may be imperfect but more sensible intuitions to the effect that in any event there is less to the distinction than meets the eye. Paul Lazarsfeld and Allen H. Barton have cited a 'direct line of logical continuity' from 'qualitative classification to the most rigorous forms of measurement' by means of intervening steps including 'rating scales, multidimensional classifications, typologies, and simple quantitative indices' (Lerner and Lasswell, 155). Here the distinction is preserved and used to establish a continuum upon which to place a series of devices: but is the continuum justified? Abraham Kaplan takes a deeper cut at the distinction by denying that there is any status to an element defined as quality or quantity apart from the choice of symbolism that has been made (Brodbeck, 602). Yet the turn from ontology to semantics is at least slowed if one wonders if any considerations influence that choice; allowing that some do could return us to the problem of ontology again.

Ernest Nagel has presented a still more pragmatic approach by arguing, first, that the advantage of numerical measurement and the reason for its success after Descartes is 'elimination of ambiguity in classification and the achievement of uniformity in practice' (Danto and Morgenbesser, 121–122); and, second, that reference to the 'real' properties of things began to appear after numerical measurements had been introduced, thus coins are considered round because of the ease of measurement thereby achieved. In reminding us about non-numerical measurement Nagel does us a service, as he does in noting that the ontological assumptions changed after the methodological conditions did. Nagel's treatment puts the question in the hands of the historian, but one may still wonder if historical trends have philosophical justification.

Finally, instead of depreciating the distinction by emphasizing the virtues of quantification one might do so by asserting the empirical priority of qualities. Something like this atittude appears to underlie May Brodbeck's treatment. In alleging that 'quantification clearly is not a condition for scientific knowledge' (574) she appears to agree with Nagel, but she is prepared to go further:

> ... the qualitative-quantiative dichotomy is spurious. Science talks about the world, that is, about the properties of and relations among things. A quantity is a quantity of something. In particular, it is a quantity of a 'quality,' that is, of a descriptive property. A quantitative property is a quality to which a number has been assigned. The colors, sounds, smells, and textures of direct experience are qualitative features of reality. So are heat and cold, round and large, heavy and solid (573).

On this account, the pragmatic standard is secondary to that of descriptive practice, but both Brodbeck and Nagel distance ontological considerations and doubt that the distinction has much merit. For Brodbeck, at least, a kind of category mistake has been committed, since talk of qualities subsumes that of quantities but not the reverse; therefore the two concepts should be seen to operate at different discursive levels.

7. QUANTIFICATION AND INTERPRETATION

Brodbeck is surely correct that the dichotomy is spurious. By way of a reminder that the field for characterizing the social world is not bipolar, recall that not all formal, axiomatized systems of description are expressed in quantities. Examples abound from the propositional calculus ('If p, then p'), to an axiomatized empirical system in which all the axioms and other propositions are given empirical value. Axiomatized systems with or without numbers are particularly at home in what David Braybrooke calls naturalistic social science, in which causal explanation is the aim (Ch. 1).

What Braybrooke calls interpretative social science, which seeks social rules, is often set up as the foil to the naturalistic sort. Falling into a dichotomous spirit, interpretivists accuse naturalists of the slavish adoration of scientific method in the physical sciences, signified by a Pythagorean number worship, or perhaps a compulsion for quantities. Thus interpretivists forget about formalism without quantification. Pressing the point further, Braybrooke tries to show that interpretative social science can – and some-times does – itself make use of quantitative methods. Significant activities on the part of social agents can be understood as the invocation of quantitative reasoning by the agents themselves as well as by the scientific observer, as in the case of the consumer deciding exactly how much he or she is willing to pay for a certain item (63). Conversely, Braybrooke defends Herbert A. Simon's mathematization of George Homans' generalization about friendship in human groups as a helpful clarification of that theory's implications (40).

While many nonquantified but mathematical formalisms have been advanced in recent years for use with qualitative data – set theory and topology principal among them (Sylvan and Glassner) – the very urge to avert imprecision remains open to question, the counter argument being:

Any form of language is *sufficiently precise* for the way of life in which it is used: *imprecision* is a claim consequent to changes in the relevant social practices and interests. There is no absolute, decontextualized criterion of precision. To press the contrary claim is to seek the change the existing way of life, not to understand it (Reason, 9, emphasis in original).

One is tempted, in assessing the development of views about quantities and qualities in social science, to reach a conclusion that is pragmatic in at least two senses. First, the kinds of signs employed by social scientists relate more to the aims of the inquiry than to some pre-systematic nature of the domain being studied. And second, the aims of inquiry have a great deal to do with the background and attitudes – that is to say *temperament* – of the investigator. If this approach were to descend into the *ad hominem*, doubtless it could be executed numerically or otherwise.

REFERENCES

Baker, Keith Michael, *Condorcet*. Chicago: University of Chicago Press, 1975.

Baker, Keith Michael, 'The Early History of the Term 'Social Science',' in *Annals of Science*, 20–, 1964.

Barton, Allen H. and Paul F. Lazarsfeld, 'Quantitative Measurement in the Social Sciences,' in *The Policy Sciences*, eds. Daniel Lerner and Harold D. Lasswell. Stanford: Stanford University Press, 1951.

Braybrooke, David, *Philosophy of Social Science*. Englewood Cliffs, NJ: Prentice-Hall, 1987.

Brodbeck, May, ed., *Readings in the Philosophy of the Social Sciences*. New York: MacMillan, 1968.

Cantore, Enrico, *Scientific Man*. New York: ISH Publishers, 1977.

Condorcet, Marquis de, *Condorcet: Selected Writings*, ed. Keith Michael Baker. Indianapolis: Bobbs-Merrill, 1976.

Freud, Sigmund, 'The Question of Lay Analysis,' *Standard Edition*.

Kaplan, Abraham, 'Measurement in Behavioral Sciences,' in *Readings in the Philosophy of the Social Sciences*, ed. May Brodbeck. New York: MacMillan, 1968.

Kiernan, Colm, *The Enlightenment and Science in Eighteenth Century France*. Geneva, Switzerland: Institut et Musée Voltaire, 1968. Second Edition, Banbury, Oxfordshire: Cheney & Sons, 1973.

Nagel, Ernest, 'Measurement,' in *Philosophy of Science*, eds., Arthur Danto and Sidney Morgenbesser. Cleveland: Meridian, 1967.

Plato, *Republic*, trans. F. M. Cornford. Oxford: Oxford University Press, 1972.

Reason, David, 'Mathematical Models and Understanding Social Life,' unpublished paper, University of Kent at Canterbury, n.d.

Reid, Thomas, *The Works*. Edinburgh: Maclachlan, Steward and Company, 1849.

Roszak, Theodore, *The Making of a Counter Culture*. Garden City, NY: Doubleday, 1969.

Sampson, R. V., *Progress in the Age of Reason*. Cambridge, MA: Harvard University Press, 1956.

Smith, Robert B., *An Introduction to Social Research*. Cambridge, MA: Ballinger, 1983.

Spradley, James, *The Ethnographic Interview*. New York: Holt, Rinehart and Winston, 1979.

Sylvan, David and Barry Glassner, *A Rationalist Methodology for the Social Sciences*. London and New York: Basil Blackwell, 1985.

PETER CAWS

THE LAW OF QUALITY AND QUANTITY,
OR WHAT NUMBERS CAN AND CAN'T DESCRIBE

ORIGINS

Before there was writing, any culture carried by language had to be transmitted orally. People memorized poems that incorporated the knowledge that
was to be passed on to future generations. A poem is something *made* (*poiein*
is 'to make'), something made with words and remembered, not just words
uttered for an occasion and forgotten. Now, we are accustomed to think,
things have changed: there are texts and chronicles, and the art of memoriza-
tion has gone almost entirely out of use. We don't need it for the storage or
transmission of knowledge, and the old chore of learning poems by heart in
school has been almost entirely dispensed with. Feats of memory, outside
some technical contexts (in the theater or in medicine, for example) have
become curiosities, useful to intellectuals who are unexpectedly imprisoned
and need something to keep them sane, but otherwise merely freakish or
decorative.

It is worth noting, though, that in fact there are still at least two poems that
everyone who has the most rudimentary education learns and remembers.
Learning them indeed is a condition for participation in the literacy that
makes the old feats of memory unnecessary. One of them is the alphabet, and
the other is the series of names for the integers.[1] They don't look like poems,
but on reflection they obviously are poems: words that belong together, to be
remembered and recited in a given but not intuitively obvious order. The
order is important, and must be learned exactly; later on it will *seem* intuitively obvious, but that will be only because it was thoroughly learned before
the concept of the obvious (or not) had been acquired.

The elements of these poems have iconic representations, in our case
respectively Roman and Arabic – a significant detail, this, and relevant to the
separation of the quantitative from other predicates in our scheme of
concepts. The Greeks and Romans used letters for numerals; in Greek they
were accented, but in both cases it was clearly enough understood that the
combinatorial rules were different as between literal and numerical uses,
whether ordinal or cardinal. We however learn different poems and not

13

Barry Glassner and Jonathan D. Moreno (eds.),
The Qualitative-Quantitative Distinction in the Social Sciences. 13–28.
© 1989 *by Kluwer Academic Publishers.*

merely different rules, so that they seem from the beginning to belong to different domains, mixing the elements of which creates awkwardness, though it is easier for us in the ordinal than the cardinal case. We may identify, and if desirable order, paragraphs, buses, telephones, postal codes, registered automobiles, etc., alphabetically or numerically or by a combination or alternation of the two, and be comfortable with this, but the alphanumeric notations sometimes used in computer programming (such as the hexadecimal, which inserts A through F between the usual 9 and 10, 1A through 1F between the usual 19 and 20, 9A through FF between 99 and 100, and so on) still seem intuitively strange to most people.

Of course it is not only alphanumeric notations that perplex – so do purely numeric ones to bases other than ten. That is because the number poem is a poem to base ten; the sequence 1, 10, 11, 100, 101, 110, 111, 1000 in the binary system would have to be read 'one, two, three, four, five, six, seven, eight,' not 'one, ten, eleven, a hundred,' etc., in order to refer correctly in ordinary language to the numbers in question, and this strains the intelligibility of the written characters. Something of the same sort happens with Roman numerals – most of us have to make a more or less conscious translation of MDCXLVII into 1647 as we read it off, much as we do with familiar words in an unfamiliar script, Cyrillic for example (try reading 'CCCP' *as* 'SSSR').

So far these considerations are purely discursive – they do not bear on the properties these two systems of representation may serve to articulate, but only on the existence of the systems themselves, and the ordering and legibility of their elements, the letters and numerals. But it is evidently not just a curiosity that these systems should exist, and it is worth reflecting on what brought them into being. Letters were the issue of a long evolution of modes of representing what could be conveyed in speech, pictorially and then pictographically and then hieroglyphically. (No doubt at the same time speech itself developed to express distinctions that had shown up graphically.) At some point the connection between the system of representation and the *content* of what was said gave way to a connection between the system of representation and the *sound* of what was said. This reinforced a separation between discourse and the world that had begun far earlier with the abandonment of any necessary connection between sound and sense, a move from motivated soundelements to merely differential ones.

With the numerals the story was somewhat different. They seem to have been invented (if etymology is to be trusted at all) in connection with a special social activity, the acquisition and distribution of goods (Latin *numerus* is connected with Greek *nemo*, to deal out, dispense, thence to hold, possess etc.; one of the derivatives of this verb is *nomos*, meaning among other things a law that assigns lots and places to people and things, from which in turn philosophers of science have derived 'nomological,' thus indirectly reinforcing the connection between mathematics and the laws of nature). This activity necessarily involved on the one hand gathering and counting, on the other dividing, apportioning and so on, and one can imagine the closeness of the attention paid to the sizes and quantities of things in these processes. The concepts of more and less are attached to powerfully affective modes of relating to the world, involving property and justice, security and self-esteem. It has been noticed by educators among others that people with apparently undeveloped mathematical talents may be quantitatively knowledgeable or even sophisticated when their interests in fair shares or sums of money is engaged.

TWO KINDS OF PREDICATE

There is an interesting difference in the uses of these two systems, hinted at above in the remark about ordinality and cardinality. Either system can be exploited for the purposes of *ordering*, on the basis of the conventional structure of its poem: we know that K comes before L just as we know that 11 comes before 12. But the development of the alphabetic system goes in the direction of arbitrary associations of letters and sequences of letters with sounds, and thence of the arbitrary association of sounds with conceptual contents, that is in the direction of language and its 'double articulation.' The development of the numeric system, on the other hand, goes in the direction of systematic combinations of numbers and thence of their systematic interrelations among themselves; in so far as they are associated with conceptual content this remains external. Words mean by referring to things in the world, numbers don't – they mean only themselves, though they can be attached to and modify the referents of associated linguistic elements. If I say 'ten grey elephants' the terms 'grey' and 'elephant' refer to each of the entities in question or to their properties, but the term 'ten' doesn't refer to any of them or even to all of them *as the entities they are*, it refers only to the cardinality of the collection to which they happen to belong. If I had said 'ten grey owls' it would make sense to ask, of 'grey', whether it was the same grey as in the case of elephants or a different grey? but it wouldn't make sense to ask if it was the same ten, or a different one.

Is 'ten' an adjectival property of 'ten grey elephants' in the same sense that 'grey' is? That, in a nutshell, is the problem of the qualitative and the quantitative. It certainly looks as if there is a radical difference here: they couldn't be elephants if they weren't grey but they could certainly be elephants if they weren't ten. Well, they couldn't be *ten* elephants, but that sounds tautological. Wait a minute, though – why not say similarly that the only thing ruled out by their not being grey is their being *grey* elephants (they might still be pink elephants)? Even so we are tempted to feel that the greyness (or pinkness as the case may be) inheres in the elephants in a way that the quality of being ten does not; going from ten to eleven is a contingent and external move, requiring nothing more exotic than the arrival of another elephant, whereas going from grey to pink seems like an essential and internal move, requiring a general metamorphosis on the part of all ten elephants.

'The quality of being ten' – this expression sounded natural enough when I used it a few lines back. It wasn't a quality of the elephants exactly, but rather of the collection they happened to constitute, which however might quite as well have been constituted by penguins, nebulae, or abstract entities. Call it a set: students of elementary abstract set theory have to get accustomed to the irrelevance of the obvious properties of the members of sets as individuals, to dealing with sets whose only members are, say {Napoleon, and the square root of minus one}, or {the empty set, and the Lincoln memorial}, and recognizing that the cardinality of these sets, which we call *two*, is the same (and the same as the cardinality of the set that contains both of *them* – and of the set that contains {the empty set, and the set that contains both of them}).

The quality of cardinality is something that only sets have; what it permits is an unambiguous classification of sets according as they have more or fewer members than, or the same number of members as, other sets. Perhaps I should have said, the *qualities* of cardinality, since two is different from three and both are different from 10^{10}. At the lower end of the scale of cardinals (it doesn't have an upper end) these qualities are *perceptible* and have common names: pair or couple, triad or threesome, etc. Other names for numbers (dozen, score) are generally survivals from alternative poems rather than directly descriptive predicates: applying them correctly normally requires counting out. In special (pathological?) cases the perception of cardinality can apparently go much higher: the neurologist Oliver Sacks recounts (an expression that in itself reflects the overlapping of the qualitative and the quantitative in ordinary language) an episode in the lives of a pair of *idiot savant* twins in which someone drops a box of matches and they both say at once '111!' When asked how they could count so quickly they say they didn't count, they *saw*:

They seemed surprised at my surprise – as if *I* were somehow blind; and John's gesture conveyed an extraordinary sense of immediate, *felt* reality. Is it possible, I said to myself, that they can somehow 'see' the properties, not in a conceptual, abstract way, but as *qualities*, felt, sensuous, in some immediate, concrete way? (1985:190).

The choice of words here reinforces the suggestion that the rest of us too think of *small* numerical attributes as qualitative, and that they become properly quantitative only when the numbers are too large to be attributed without counting.

If we now revert from speaking of sets as such to speaking of their members, we say that there is a *quantity* of them – but they don't thereby acquire any new *qualities*. Things get complicated, though, when this habit of switching attention back and forth from sets to members of sets follows the development of the number system from the integers or natural numbers, in connection with which the idea of cardinality was first defined, to rational, real, or even complex numbers. By introducing the concept of *unit*, which makes some standard embodiment of a quality such as length or weight (the standard meter, the standard kilogram) the sole member of a set of cardinality *one*, and specifying a rule of matching (laying end to end, piling up in the scale of a balance) that will generate sets of higher cardinality whose members will be units (fractions of units being relegated to fractional scales, where the new units are fractions of the old, a tenth, a hundredth etc.), cardinality comes to be attached by courtesy to other objects embodying the quality in different degrees. Instead of 'longer' and 'shorter' we now have '11 meters' and '10.3 meters,' which define whole classes of longers and shorters among indefinitely many such possible classes. Our *interest* in '11 meters' as a defining property of some object was initially, no doubt, a desire to know what it was longer or shorter than, or the same size as, but '11' came to attach to it as a predicate along with 'blue,' 'soft,' 'glutinous' and whatever other qualities our postulated 11-meter object may be supposed to have. And before we knew it our language was stocked with ratios, averages, angles, temperatures, coefficients, dates, times, indices, prices, and other numerically-expressed predicates as familiar and useful, in our commerce with things in the world, as any other qualities by which they might be distinguished from one another.

THE LAW OF QUANTITY AND QUALITY

The specifications of degree among objects sharing a given quality, that quantitative predicates make possible, have been available in some technical contexts for a long time, but their general invasion of daily language is relatively recent. That the temperature should be 'in the sixties' has of course been a possible determination only since the invention of the Fahrenheit scale and the general availability of thermometers, i.e. since the early eighteenth century. But a temperature 'in the sixties' has nothing to do with the number 60 or the cardinality it represents, it has to do with spring and light coats, while 'in the twenties' means bitter cold and 'in the nineties' intolerable heat. Note that the expressions 'in her twenties,' 'in his nineties,' coexist with these unambiguously, as indeed do 'in the twenties,' 'in the sixties,' and so on as applied to the years in a given century, but that these in their turn mean young and beautiful or old and wizened, flappers and flower children, rather than anything quantitative. It is interesting to find that although these latter expressions have been available for much longer than is the case with the weather, birthdays and calendars having been marked by cardinals for centuries, they were not in fact used until about the same time; whether it be temperatures, ages, or years, the first occurrences of the expressions 'twenties,' 'thirties,' etc., up to 'nineties,' are all given by the Oxford English Dictionary as falling between 1865 and 1885.

It was at about this time, in 1878 to be exact, that Frederick Engels, in *Herr Eugen Duhring's Revolution in Science* (commonly known as the 'Anti-Duhring'), gave popular form to the principle, introduced by Hegel and utilized by Marx, of the passage of quantity into quality. Hegel speaks of 'nodal lines' in nature, along which incremental quantitative changes are accompanied, at the nodes, by qualitative shifts. Such a shift is 'a sudden revulsion of quantity into quality,' and Hegel offers as an example 'the qualitatively different states of aggregation water exhibits under increase or diminution of temperature' (1975:160). Engels too cites this as 'one of the best-known examples – that of the change of the state of water, which under normal atmospheric pressure changes at 0°C from the liquid into the solid state, and at 100°C from the liquid into the gaseous state, so that at both these turning-points the merely quantitative change of temperature brings about a qualitative change in the condition of the water' (1939:138).

This 'brings about' however is highly misleading. It gives the impression that temperature is a property of water that is causally related to its state: change the (quantitative) temperature, and the (qualitative) state will change.

The fact is that at the boiling and freezing points the temperature *can't* be changed *until* the state has changed. What happens is this (I will take the case of boiling, which applies *mutatis mutandis* to freezing also): steadily supplying enough heat energy to water will raise its temperature to 100°C; at this point supplying further energy will not change the temperature but will dissociate the molecules from one another so that they become steam at 100°C; when all the water has been changed to steam then, assuming a closed system, the supply of still further energy will raise the temperature of the steam above 100°C. But if the process begins at room temperature it will take about seven times as long to change all the water into steam as it took to raise the water to the boiling point.

So there are two things wrong with the Hegel–Engels account: first, it isn't changing the temperature that changes the state, and second, the change is not sudden. As I have pointed out elsewhere (1972:78), when water boils because it is heated from the bottom, the change of a small amount of it into steam makes dramatic bubbles, and this is not a bad analogy for *repressed* change, which was one of the popular senses in which the dialectical principle of quantity and quality came to be understood: history will accumulate exploitation and repression incrementally, until crisis and revolution suddenly ensue. And this may indeed happen – only the quantity/quality distinction has nothing to do with it. Water froze and boiled long before temperatures were thought of, and when we talk about 'the boiling point' and attach a number to it (note by the way that it is impossible to *measure* the boiling point at standard atmospheric pressure in degrees Celsius, since 100°C is *defined* as the boiling point of water at standard atmospheric pressure) the number by itself does not refer to anything that is true of the water, but (as before) only to the cardinality of a collection of units.

This point can be driven home in various ways. One of the remarkable and useful features of the exact sciences is that quantities can be measured and the measurements plugged into computations. The qualities whose degrees are attended to in the process of measurement (or predicted by the outcome of the computation) are sometimes thought to enter into the computations. Thus in the most elementary case of a freely falling body initially at rest we have the equation:

$$s = 1/2 \, g \, t^2,$$

which means 'the distance fallen is equal to half the acceleration of gravity multiplied by the square of the time elapsed.' But a moment's thought will show that this can't possibly be what is meant: times can't be squared, only numbers can; nothing can be multiplied by an acceleration. The expression is only a shorthand way of saying that measurements of the distance, the acceleration, and the time, using compatible units, will yield numbers that stand in the required arithmetical relation. In the algebraic expression given above s isn't a distance at all, it's a variable that can take numerical values, and so for the other elements.

The coincidence of Engels' popularization of dialectical doctrine on the one hand, and the emergence of numerical expressions as descriptive in ordinary language on the other, suggests that the latter paved the way for the general confusion represented by the former. We can use numbers to describe things, but unless the thing described is a set or collection with a given cardinality they won't be functioning *as* numbers, just as predicates to be defined in the ordinary way and eliminable by substitution. Their use will be a metaphorical use. Yet in the last hundred years or so people have thought of themselves as getting hold of a special numerical or even mathematical feature of things when they use numbers in this way, a quantitative feature at any rate. And when the numbers change concomitantly with some notable qualitative change we have all the appearances of a passage from quantity to quality.

NOTABLE AND JUST NOTICEABLE DIFFERENCES

The idea of concomitant change ('concomitant variation,' to use Mill's phrase) is basic to the scientific enterprise: we want to know, if we make some change in the world, what else will also change, so that we can achieve or avoid it. Changes can be large or small, dramatic or marginal. Group sizes change by the addition or subtraction of members, other properties by augmentation or diminution, intensification or dilution, etc., or by outright metamorphosis, one property being replaced by another. Cumulative marginal changes, each of which is hardly noticed, may eventually result in states so altered that they require altogether different descriptions. But this phenomenon is context-dependent and works on both sides of the qualitative-quantitative boundary. If a large surface, a wall for example, has always been red, but suddenly overnight is painted yellow, the change is startlingly obvious, but if its red color is modified very slowly, through an imperceptible shift in the direction of orange and progressively through lighter and lighter shades, until finally the last trace of red has vanished and the wall is pure yellow, the fact that it has changed at all may dawn only slowly, and then only on an observant witness with a good memory (imagine the change stretched out over centuries, so that in any one witness's life it was just an orange wall). Psychologists speak of 'jnd's' or 'just noticeable differences' as a measure of the refinement of perception (similar to 'resolving power' in optics), a threshold below which changes cannot be perceived, so that several subliminal moves may be possible before anything is noticed – and indeed if they are made at suitable intervals nothing may ever be noticed.

Something very similar happens on the quantitative side if the sets in question are sufficiently large. If one person is in a room and another enters, the change is obvious enough, and similarly if a third joins a couple, but if forty people are watching a parade, let us say, the arrival of the forty-first may go entirely unremarked. Still if people keep coming, one by one, sooner or later we have a huge crowd, a demonstration, a triumph – and when exactly did this happen? There is an ancient paradox called The Heap: a grain of wheat is set down, then another grain, and so on; eventually there is a heap, but which grain was it that turned a scattering of grain into a heap? This paradox was presumably intended to remain paradoxical – no empirical research was done, as far as I know, to find out when impartial observes would start to use the term 'heap' without prompting. (My guess is that four grains, in a tight tetrahedral array, would qualify as a very small heap, whereas if the procedure were to scatter randomly over a given area, say a square yard, there would be a range of many thousands of grains over which the status of the accumulation as a heap could be disputed.) The point the paradox makes is that categoreal boundaries, for example between 'scattering' and 'heap,' are fuzzy, but that surely comes as no surprise and hardly makes a very convincing foundation for philosophical doctrine, whether metaphysical or revolutionary.

The dialectical law of the passage of quantity into quality, like its companions, the law of the interpenetration of opposites and the law of the negation of the negation, is thus seen to be an entertaining but non-essential red herring. There are cases in which cumulative imperceptible changes in x lead to the emergence of y, and there are cases in which they just lead to more x – and either x or y can be indifferently qualitative or quantitative predicates; everything depends on the particular case, and can only be learned by looking. Adding atom after atom to a lump of uranium 235 eventually produces an atomic explosion and an assortment of vaporized fission products; adding atom after atom to a lump of gold just produces a bigger lump of gold. Water when refrigerated changes into ice, iron when refrigerated gets colder but doesn't change into another form. No general law can be established that would be of any reliable predictive value; as in any empirical situation the correlations cannot be generalized in advance but must be learned for each case or class of cases. That solids will eventually melt on heating, and liquids vaporize, can be expected within limits, but even there other forms of dissociation may take place, and nothing whatever is gained by claiming these phenomena as examples of the dialectic in nature.

The contingency of the relation between quantitative and qualitative change, its dependence on the state of the system, can be illustrated by the following thought-experiment, in which A is a pedestrian walking slowly towards the edge of a cliff C:

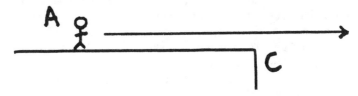

Cumulative quantitative displacements of A in the direction of the arrow will lead to a dramatic qualitative change in his situation at point C (call it the 'falling point'), but nobody would seriously think of attributing this to the quantitative change as such, only to its taking place near the edge of the cliff.

These considerations do not abolish the differences between qualitative and quantitative but they do suggest fresh ways of thinking about them. In particular it is not clear that they need be accepted as dividing the field when it comes to determinations of the state of the world in various respects. Both derive from members of a family of Latin adverbs beginning with 'qu-,' all of which have interrogative uses, whose form was presumably determined by the verb *quaero*, to seek, ask, inquire. So *qualis*? from which 'qualitative' derives, means in effect: 'I ask: what sort?' while *quantus*? similarly means 'I ask: how much?' We may think of this 'qu-' prefix as a kind of question mark, and translate *qualis* and *quantus* respectively as '(?)sort' and '(?)degree.' However there are lots of other possible questions, and Latin provides for them: (?)manner will give *quam* or *quomodo*; (?)time, *quando'* (?)elapsed time, *quamdiu*; (?)reason, *quia* or *quare*; (?)distance, *quoad*; (?)place, *quo*; (?)number, *quot*; (?)frequency, *quoties*; (?)number in series, *quotus*, and so on. Why should there not therefore be quamitative, quanditative, quaritative, quotative and quotitative inquiries, as well as qualitative and quantitative? And yet these last two are the only survivors to have made it into our ordinary language, and this means, if we are to take Austin seriously, that only one difference or opposition out of this whole crew was important enough to be preserved. The question is, what opposition was it?

I shall suggest that it was *not* the sort of opposition that divides the world into a part that is qualitative and a part that is quantitative, or that allows the transition of one sort of predicate into the other according to any law, no matter how dialectical. The world is as it is and its states are amenable to description on condition of our having a suitable language at our disposal; every element of every state invites the question what sort of thing it is, what sort of thing is going on. Let this be the general question, the descendant of *qualis?* to which the answer may be in diverse modes, spatial, temporal, causal, numerical and so on. If the last among these rather than the others singles itself out for special attention, why might this be?

'SEPARATION OF THE MATHEMATICAL APPARATUS'

It should be noticed at once that something is slipping here – if *numerical* properties had been the issue surely *quot* rather than *quantus* should have been the root of our own expression. This slippage indicates, I think, where our own confusion lies. The questions 'what sort?' and 'how much?' are both required if the entity or event under investigation is to be estimated correctly in relation to other things; both are *differential* questions, and the answers to them provide the coordinates that locate the object in an array of types and magnitudes: the first distinguishes it from other objects of different sorts, the second compares it with other objects of the same sort. The latter purpose, however, can be served in diverse ways – within a given category there can be more than one dimension of variety. So a series of possible orders may be envisaged, in which the members of the category might be arranged, and for each order an ordinal sign may be assigned to each member. For this purpose we are not unlikely to call upon one of the poems we began with. And the discovery that if we choose the number poem we may also be able to make use of cardinality, and even perform computations that will accurately predict some features of the ordering in question, will come as a surprise and a revelation.

It is just this formal and computational aspect of the matter that brings in the quantitative as it has generally come to be understood. One of the earliest discoveries along these lines was made by the Pythagoreans, who correlated the ratios of lengths of stretched strings with musical intervals. They thought this discovery sacred, and indeed it is hard to imagine the awe and astonishment it must have produced. I suspect (indeed I remember) that something like it can happen in childhood when elementary mathematical truths suddenly dawn, but that is an expected step, an entry into a known domain, not as for them the opening up of something novel and incredible. Pythagorean doctrine concluded that the world was at bottom *numerical*, which involved a category mistake but nevertheless set the tone for a long tradition. The beginning of modern science was marked by Galileo's resolve to make the 'definition of accelerated motion [i.e. its mathematical expression] exhibit the essential features of observed accelerated motions' (1914:160), a scrupulous formulation that seems unnecessary to us, because obvious, but that required new clarity on his part. The comparable claim in his case was that 'the book of nature is written in the language of mathematics,' which does not involve a category mistake but does assume a parallel between an intelligible domain (the book and its mathematics) and a sensible one (nature); here also Galileo was scrupulous and clear, though his remark has frequently been interpreted as meaning that 'nature is mathematical,' which brings back the mistake.

These episodes represent steps in a process of realization that reached its full formulation with the Turing machine: the realization that *all* relations between exactly specifiable properties of all the things in the world can be modeled to as close an approximation as desired in logico-mathematical language. This development is recounted with great perspicuity in Husserl's *The Crisis of European Sciences and Transcendental Phenomology*, in which he speaks of 'Galileo's mathematization of nature,' and in a brilliant image describes a tendency to 'measure the life-world – the world constantly given to us as actual in our concrete world-life – for a well-fitting ... garb of symbols of the symbolic mathematical theories' (1970:51). The success of this program of measurement however leads to 'the surreptitious substitution of the mathematically substructed world of idealities for the only real world, the one that is actually given through perception, that is ever experienced and experienceable – our everyday life-world' (1970:48–49).

The properties of things in the life-world are what we would normally and generally call 'qualities,' and the only qualities that permit of direct mathematical expression are precisely the properties of sets or collections already discussed; all the others have to be translated into sets or collections, through the specification of units and combinatorial procedures. This process has been called 'substruction' by Paul Lazarsfeld, independently, I take it, of Husserl's use of the term (cf. the quotation above); it 'consists essentially in discovering or constructing a small number of dimensions, or variables, that underlie a set of qualitative types' (Selvin 1979:234). The actual carrying out of the process will involve distinctions between ranked and scalar variables, discontinuous and continuous scales, ratio and interval scales, etc. (see Ghiselli 1981:12ff.); fitting the life-world with its mathematical garb is a busy and demanding industry.

Only sets or collections, properly speaking, can be said to have quantitative properties, and these in the end will all turn out to be numerical – Husserl speaks of the 'arithmetization of geometry,' of the transformation of geometrical intuitions into 'pure numerical configurations' (1970:44). (This claim is no doubt oversimplified – there may be topological features, such as inclusion or intersection, that have non-numerical expressions, though these would not normally be called quantitative.) What are thought of as quantitative properties of other entities, such as length, temperature, density etc., are so many qualitative properties with respect to which however an entity may change its state over time, or otherwise similar entities may differ from one another. Such differences are themselves qualitative, though they may be given numerical expression. It is important to realize that, for example, the difference in height between someone five feet tall and someone six feet tall is not a numerical difference, even though the difference between five and six is a numerical difference. At every given instant every entity is in the state it is in, with the qualities it has. These may include vectors of change or becoming. Whether such vectors essentially involve quantities – that is, whether becoming involves at every infinitesimal moment a change in the size of a collection – is a question as old as Zeno, which however need not be answered in order to characterize a momentary state.

Of course collections may change their cardinality with time, and we can over suitably large time intervals *make* other changes into changes in the cardinality of collections by choosing to represent them numerically. In counting and measuring we have two ways of generating numerical predicates out of determinate qualitative situations. The numbers so generated can be inserted into more or less complicated mathematical expressions and made the objects of computation; the numerical outcome of the computation may then by a reverse process be applied to a new qualitative feature of the original situation, or to the same feature of a transformed situation. The rules according to which all this is done (the generation of the numbers, the computations, and the application of the results) have to be learned empirically, as Galileo realized; in this way a number of mathematical relations and formulae are selected from the potentially infinite store of such things and given physical meaning by courtesy. But the mathematical work is entirely carried out within mathematics; measurement shifts attention from quality to quantity, crossing the boundary between the sensible and the symbolic. This shift corresponds to what Braithwaite, in his *Scientific Explanation*, called the 'separation of the mathematical apparatus' (1953:47–49).

QUALITATIVE AND QUANTITATIVE REVISITED

Qualitative and quantitative do not divide up a territory, they both cover it, overlapping almost totally. But one is basic and the other optional. Everything in our world is qualitative; but virtually everything is capable – given suitable ingenuity on our part – of generating quantitative determinations. Whether we want to expend our ingenuity in this way is up to us. The United States Bureau of the Census, whose main business might seem to be quantitative, has nevertheless an interest in questions of 'the quality of life,' and has devoted a good deal of attention to efforts that have been made to translate expressions of satisfaction or dissatisfaction into numerical measures. The standard trick is to develop an ordinal ranking and then assign cardinal values to the positions within it for the purpose of drawing graphs, performing statistical computations, etc. The SIWB scale, for example (the initials stand for Social Indicators of Well-Being), assigns the integers 1 through 7 to 'terrible,' 'unhappy,' 'mostly dissatisfied,' 'mixed,' 'mostly satisfied,' 'mostly pleased,' and 'delighted' (Johnston 1981:2).

One possible use of the results of inquiries on such bases (or improved ones – the Census people seem realistically aware of the shortcomings in the state of their art) might be to produce correlations between these measures and quantities that permit of objective assessment, such as income, energy consumption, cubic feet of living space, number and horsepower of automobiles, etc. These might throw light on some aspects of our common systems of value. But it is worth noting that the starting-point here is not an experimental procedure but an appeal to the judgment of an individual. The individual does not need the quantitative apparatus, only in the first instance an awareness that better or worse conditions are possible, and a subjective conviction of distress or euphoria as the case may be. This is what I mean by saying that the quantitative is optional: our lives would be in some important respects just what they are if we did not know the date or the time or the temperature, or perhaps even our ages or bank balances or IQ's or cholesterol counts. In some significant respects they might be better. I do not mean this as a regressive criticism of measurement or computation, without which we would be at the mercy of old forces from which they have helped to deliver us, but rather as a comment on the use of the metaphorical language of number.

The French used to make fun of tourists who insistently wanted to know the population of this city, the height of that building, by calling them *hommes chiffres*, 'number people.' It is worth asking what use is to be made of numerical information. Sometimes numbers are reassuring or threatening, as when they mean that I can expect to live a long time, or that I run such and such a risk of having a certain sort of accident. Sometimes they give me a sense of solidarity with a community, sometimes a sense of inferiority or superiority. Sometimes there is an effect of scale, as when the numbers of people killed at Hiroshima or in the Holocaust boggle the imagination – genuine cases, perhaps, of a *psychological* transformation of quantity into quality (and with nothing metaphorical about the numbers either). But in every case, even these, I or other individuals must prosper or suffer singly. The quality of pain or terror or despair involved in a quite private injury or death or betrayal may match anything any individual can feel or have felt in a mass event.

The value of genuinely collective measures – aggregates, averages, and the like – remains unquestioned, but the question as to role of numerical determinations in the descriptive vocabulary remains open. Part of my argument has been that when these come about as a result of measurements they are to be understood not as quantities but as disguised qualities. Their use as such has drawbacks as well as advantages. There is a short story of Hemingway's, 'A Day's Wait,' that may serve as a closing illustration. An American child who has lived in France falls ill, and overhears the doctor telling his father that he has a temperature of 102°, upon which he withdraws into himself, stares at the foot of the bed, and won't let people near him. Only at the end of the day does it dawn on his father that he takes this 102 to be in degrees Celsius, a scale on which he has been led to believe a temperature of 44° to be surely fatal, and that he has been quietly preparing for death. The story ends on a happy if a shaky note. But in a world where plunges in the stock market index have been known to provoke plunges from high windows there may be room for the renewed cultivation of quality unmediated by quantity, leaving the quantities to do their undeniably useful work in their proper domain.

NOTE

[1] There are two other poems of the same sort that everyone also learns – the names of the months, and of the days of the week – but these do not open up comparably large domains of conjecture and argument.

28   PETER CAWS

REFERENCES

Braithwaite, Richard Bevan, 1953, *Scientific Explanation: A Study of the Function of Theory, Probability and Law in Science*. Cambridge: at the University Press.

Caws, Peter, 1972, 'Reform and Revolution,' in Held, Virginia *et al.*, eds., *Philosophy and Political Action*. New York: Oxford University Press.

Engels, Frederick, 1939, *Herr Eugen Duhring's Revolution in Science*, tr. Emile Burns. New York: International Publishers.

Galilei, Galileo, 1914, *Dialogues Concerning Two New Sciences*, tr. H. Crew and A. De Salvio. New York: MacMillan.

Ghiselli, Edwin E., Campbell, John P., and Zedeck, Sheldon, 1981, *Measurement Theory for the Behavioral Sciences*. San Francisco: W. H. Freeman and Co.

Hegel, G. W. F., 1975, *Hegel's Logic, Being Part I of the Encyclopedia of the Philosophical Sciences ((1830) translated by William Wallace)*. Oxford: at the Clarendon Press.

Husserl, Edmund, 1970, *The Crisis of European Sciences and Transcendental Phenomenology: An Introduction to Phenomenological Philosophy*, tr. David Carr. Evanston: Northwestern University Press.

Johnston, Denis F., ed., 1981, *Measurement of Subjective Phenomena*. Washington: U. S. Department of Commerce [Special Demographic Analyses].

Sacks, Oliver, 1985, *The Man Who Mistook His Wife for a Hat, and Other Clinical Tales*. New York: Summit Books.

Selvin, Hanan, 1979, 'On Following in Someone's Footsteps: Two Examples of Lazarsfeldian Methodology,' in Merton, Robert K., Coleman, James S., and Rossi, Peter S., eds., *Qualitative and Quantitative Social Research: Papers in Honor of Paul S. Lazarsfeld*. New York: The Free Press.

CHARLES W. SMITH

THE QUALITATIVE SIGNIFICANCE OF
QUANTITATIVE REPRESENTATION

Few questions have the capacity to provoke as intense an emotional response among social scientists as that dealing with the utility of quantitative methods for explaining human/social behavior.[1] On the one hand, there are those who argue that only through the application of quantitative measurements and methods can the social sciences ever hope to become 'real' sciences; on the other hand, there are those who claim that the subject matter of the social sciences is simply not amenable to quantification and all attempts to impose such measures and methods upon social behavior is just so much nonsense. What makes this situation somewhat puzzling is that in most cases, each side presents the opposing side, or at least the more sophisticated spokesmen for the other side, in what can only be called caricature form.

Despite such misrepresentation, it would be wrong to conclude that there are no real differences; there are. Whereas the quantitative camp is generally willing to admit the quantification must be done with care, that there is a need to insure that one has properly isolated significant 'variables' and that these variables are amenable to quantification, etc., they still believe that the social sciences should move toward greater quantification. Alternatively, whereas members of the more intrepretive/qualitative camp are generally willing to admit that there are times when quantification may be useful and appropriate, they tend to be critical of most attempts to impose quantitative measurements and methods upon social behavior. Despite laudable efforts on the part of members from both camps to clarify the central questions, for most practitioners the situation could be characterized as one of conflicting claims: 'Yes, we can!', 'No you can't!'

If one was to conclude in light of what was just said that I am less than enthralled by most of the ongoing discussion around the qualitative/quantitative issue, she would be correct. I still find that I enjoy rereading Cicourel's *Method and Measurement in Sociology* as well as a number of pieces by Lazarsfeld, Arrow, and others, but on the whole I find most quantitative versus qualitative discussions – the articles in this volume being exceptions – somewhat sterile.[2] When all is said and done it seems to be the case that while most parameters allow in principle for quantification, in most cases our parameters are insufficiently articulated to make such attempts

29

Barry Glassner and Jonathan D. Moreno (eds.),
The Qualitative-Quantitative Distinction in the Social Sciences. 29–42.
© 1989 *by Kluwer Academic Publishers.*

sensible. It also seems to be the case that while certain aspects of much observational data can be quantified, there is often no significant payoff to doing so while in other cases attempts to impose quantitative measures on data can actually distort the material. In short when it comes to dealing with parameters and variables, the best advice seems to be that it may be possible to quantify, but in most cases it doesn't seem to be wise and may actually be counterproductive. (When I first began to do empirical research, Everett Hughes shared with me the wisdom he had acquired through years of doing empirical research, namely that (1) the problem is not how to measure and quantify, but knowing when to measure and quantify, and (2) that often it never becomes useful to measure and quantify.)

The quantitative/qualitative issue takes on a somewhat different face when placed within the context of explanation rather than the proper representation of data. Here the issue becomes more that of the utility of mathematical modeling rather than the use of quantitative data *per se*, though for many the issues are closely tied. Again the most sophisticated views seem to be that the social sciences have nothing to fear from mathematics, but on the other hand there doesn't seem presently to be much use for such explanations either. While mathematical reasoning has historically served to define and clarify issues when applied to underlying theoretical assumptions, it has also been used to obfuscate and mislead when used to subsume empirical findings which do not reflect any mathematical order. Here again, however, it seems to be a case of an awful lot of heat with relatively little light.

The quantitative/qualitative debate within the social sciences is not, of course, unique in generating more heat than light; to a large extent the same accusation could be made of most methodological/philosophical debates within the social sciences. By and large a good deal of the problem has been due to bad philosophy of science. Fortunately, it looks like some remedies are in the offing; I am here thinking specifically of the works of Roy Bhaskar which are so meaningfully applied to the quantitative/qualitative issue by Peter Manicas in his paper in this volume.[3] It really does seem to be the case of the social scientists getting their scientific model wrong and then trying very hard to make their own efforts fit the faulty model. To a large extent, I think this really is what has happened. It clearly isn't, however, the whole story. I say it clearly isn't the whole story, because what might be called the pursuit of the quantitative grail and the power of scientific prediction which is supposed to accompany it has continually been revealed as a basically fallacious enterprise by significant social theorists, even if not in quite the elegant manner employed by Bhaskar and company. In fact it seems to be a perfect example of 'when prophesy fails,' i.e., the more the inadequacies of the positivistic paradigm are revealed the more ardent becomes the commitment of its supporters.

While a good deal of the infatuation with quantitative data has probably been due to faulty philosophic models and a good deal of the continuing commitment to such modeling has probably been due to various psychological defense mechanisms, as a sociologist I still find a need for some sort of sociological explanation. Put slightly differently, I am interested in determining the sociological significance and meaning of quantitative representations on the part of social scientists. It is to this issue, therefore, that I will now turn leaving the more philosophic questions to my philosophically concerned colleagues.

To put the matter quite bluntly, I am really not interested in the question of whether or not sociological data can be quantified, or even if it should be quantified. Some of it clearly can be, but as noted earlier I believe that seldom if ever is it necessary or even particularly useful. The more interesting question is why sociologists have continued to make the effort. To answer this question, I would suggest that we look at the process whereby we attempt to impose quantification upon phenomena in general. In short, I suggest that we look at the issue as a problem in the sociology of knowledge.

Though we tend to look upon any attempt to quantify data as due to our 'scientific bent', the fact of the matter is that the 'scientific' push toward quantification represents only a small element of a much broader movement. The tendency to count and measure pervades our lives, though clearly there are times and places where it is more prevalent. We count minutes, hours, days, weeks, months and years; we measure miles traveled, sizes of lots, and miles to the planets and stars. We not only count our change and attempt to balance our check books, but we assign numbers to dives, acrobatic routines, movies, and pulchritude (the perfect 10). In short, the act of quantifying is clearly not limited to scientific pursuits.

Not only does the tendency to quantify pervade our daily lives, it appears to have also pervaded the lives of many of our forebears. One need only skim through a bible to see this biblical concern with quantification. We are not told that Seth, Enosh, Kenan, Mahalalel, Jared, Enoch, and Methuselah lived to be old men, but are given exact ages. We are given similar exact quantities when it comes to describing Noah's ark, the building of the Tabernacles and the number of persons in the various tribes. It isn't by accident that the third book is called the book of Numbers, though the other four books have their share of quantities.

It may be argued that this concern with numbers is, in the context of historical documents, peculiar to the bible. The fact of the matter is, however, that history, at least in its western form, is embedded in numbers, namely, dates. Admittedly, many historians continually attempt to free history of its calendar straightjacket, but the tendency, if not the need, to define and deal with events in terms of their numerical dates remain.

It would be possible to give example after example of our general tendency to quantify, but to do so I believe is unnecessary. The fact is that we have become – the question of whether we always were is an issue I do not want to tackle here – a counting species. We count time and space, calories, cracks on the pavement, etc., etc. What is also clear is that most of the time we are not counting in order to better understand and/or gasp whatever we are counting. So what are we doing?

While I would not want to claim that we are always doing the same thing, I think that what we most often are doing is attempting to objectify that with which we are dealing. We are attempting to give it an independent existence from our experience of it.

To a large extent, of course, this is exactly what the proponents of quantitative scientific methods claim. They argue that it is only when we can measure things can we claim to know the thing objectively and given that the goal of science is objective knowledge, science requires quantification. What I am suggesting, however, is not that quantification is the process whereby we grasp what is 'objectively' there, but rather the process, or minimally a process, whereby we intend to objectify phenomena. Put slightly differently, quantification is part of a particular way of defining the situation. It cannot consequently be understood in and of itself, but must rather be put in the context of multiple interpretive frames of reference.

Like the qualitative/quantitative issue, the subject of multiple frames of reference/experience is a 'golden oldie' which has been analysed extensively by numerous theorists. While various different schemata have been suggested by these various theorists, I have for a number of years been most comfortable working with four modalities, which I have labelled (a) the organism/libido mode; (b) the other persons/power mode; (c) the physical objects/instrumental mode; and (d) the symbolic/ordering mode. In a number of earlier pieces I have tried to show how these different modalities are grounded in different types of human activities and entail different ways of grasping and symbolically representing such activities.[4] More specifically, I have tried to show how they differ, among other ways, in (1) their foci of concern; (2) the aspects and/or dimensions of things highlighted; (3) the way things are seen as related; and (4) the degree to which subjective/objective distinctions and distances are assumed. It is specifically the last point which bears on the present discussion. I will attempt, therefore, to limit my discussion to this point, though I fear some may find my approach somewhat convoluted.

I think it is fair to say that there is presently general agreement that the essence of human reasoning is its inherently reflexive character.[5] Human reasoning is grounded in the comparatively unique ability of the species to reflect back upon its own experiences which, in turn, allows for experiences to be linked in an active manner rather than merely in the orders in which they are passively encountered. It is this reflexive process which transforms the stream of experience into our conceptions of both our selves and our worlds. What is not as generally recognized is that the way in which this reflexive process operates and the form of both selves and worlds generated differ depending upon the nature of the experience, or more accurately the dominant praxis of the experience.

When the activity is governed primarily by organismic needs, the experience is generally co-opted by the organism with relatively little attention given to the task of using the experience to define an external world. Insofar as 'objects' are generated from such experiences they tend to be labeled as 'good/pleasurable/etc.' or 'bad/painful/etc.' When the activity is governed primarily by interpersonal power objectives, there is similarly little attention given to defining the external world in a non-self oriented manner. Our concern rather is to reflect the extent to which the 'other' is stronger or weaker than ourselves. To use some common mathematical notations, we could say that the first modality tends to define the world – or more accurately experiences – in terms of positive (+) and negative (–) sets whereas the second modality tends to relate to the world in terms of greater-than, less-than relationship. It is only when we begin to experience the world as independent and external loci of experiences that we begin to make extensive use of number and quantities, though as I will attempt to show once such numbers are introduced they can be appropriated by other modalities.

Why quantification and numbers should be associated with experiences of an external physical world is a question of sufficient complexity that I cannot hope to answer it in any sort of definitive manner in the space presently available. I will attempt, therefore, simply to note some of the general factors at work. I might just add that though the following comments do not assume any set of necessary categories of experience, they are closely related to Kant's discussion of this issue.[6]

The concepts of quantfication and number are based upon the notion of equivalencies. There must be something, or rather some things, to count. This, in turn, generally entails locating such 'things' within a spatial matrix. It is, in fact, the experience of such 'objects' within a spacial matrix which most commonly allows for equivalencies. The notions of the conservation of matter and quantities require a world of external objects. To be able to count it is necessary to be able to stand off from what is being counted and it is specifically things experienced as located within a spatial-temporal matrix that we most naturally are able to do this. I am not arguing that such counting is limited to such objects, but only that counting appears to emerge from a developmental perspective within such contexts. (Piaget's research on quantification and number would seem to support this view.)[7]

The above suggested affinity between quantification and the world of objects is, of course, anything but original. It is specifically this affinity which motivates some to equate the quantification of data with the establishment of the 'objectivity' of the data. It is important to recognize, however, that the relationship is just the reverse. It is the world of objects which lends itself more readily to quantification and counting than do other modalities of experience, rather than the notions of quantification and counting which transform experiences into 'objects.' The natural affinity between quantification and experiences of objects within a spatial-temporal matrix is such, however, that we often tend to believe that the reverse relationship does hold, i.e., by quantifying experiences we somehow transform experiences into objects. To put this somewhat differently, the experience of 'objectivity' is initially grounded in experiences of spatial-temporal objects; it is specifically such experiences, however, which lend themselves to quantification with the result that the process of quantification itself acquires this sense of 'objectifying.' In the case of quantification *per se* this sense of 'objectifying' rest primarily upon the self-distancing aspect of the quantification process.

Given that many types of experiences do not naturally lend themselves to quantification, the question arises why we should attempt to quantify such experiences. Why attempt to quantify beauty or pleasure or the relative differences within an ordinal ranking? It is here that I think we begin to get at the heart of the matter. As just noted quantification carries with it the sense of 'objectively given' acquired from its normal association with experiences of spatial-temporal objects. Put slightly differently, though quantification in and of itself in no way entails 'objectivity' we often assume such 'objectivity' through association. The reason we quantify is that it conveys this sense of 'objectivity' or more correctly a sense of independently given, i.e., a sense of 'reality.'

Here we introduce another issue of some complexity, namely the relation-ship between 'objectivity' and 'reality.' In today's world the very notion of 'objectivity' seems to imply 'reality.' This linkage is of great importance since from a philosophic perspective the attraction of quantification is not simply that it entails greater objectivity, but also that such objectivity entails a better grasp on reality. The relationship between 'objective' and 'real', however, at least as the notion of objective is generally understood, namely, as empirical, is anything but certain. For Kant, for example, the world of objects, while clearly the focus of science, remains the world of appearances and not the world of fundamental reality. The paradox here is that while quantification may indeed be the preferred and desired means for understand-ing our experiences of the spatial temporal world of objects, it does not follow that quantification would better allow us to understand other types of experiences. It is not enough to say that quantification seems to entail a sense of objectivity, it is also necessary to understand how this sense of objectivity coopted our sense of reality, or how in some manner quantification itself managed to project a sense of reality. In either case the issue is to discover how quantification acquired its reputation as the universally preferred method of grasping reality?

There are a number of possible answers to this last question. Perhaps the most obvious would be that despite what Kant himself said on the matter we have come to accept the world of Kantian appearances as the world of reality. That is, we now generally accept the world of empirically experienced objects as the real world and consequently assume that the methods used to grasp these types of experiences are the best for grasping reality in general. While there is clearly some truth to this position, I fear that it is again a case of at least partially putting the cart before the horse. The fact of the matter is that from a number of different perspectives the world of experienced objects is a poor candidate for 'reality.' There is simply too much discrepancy in the ways we experience such objects. We also have other types of experiences which seem more intense and more 'real.' The history of philosophy is replete with arguments against accepting as reality what we call the objective world. Somewhat ironically, however, it has commonly been our notions of quantification and number – which as shown above emerged from such experiences – that have been proposed as alternative paths to reality. What I am suggesting here is that while a case can be made for the position that we tend to equate the quantifiable with the real because we also tend to equate the spatial-temporal world of objects with reality, it can also be shown that to be quantifiable conveys a sense of reality in its own right.

To make sense out of this situation, I would suggest that it is necessary to take a closer look at what purpose or purposes these various accounts serve.

The most obvious purpose of such accounts it to explain. The accounts are presented as attempts to enable us to understand. We seek to know. Unfortunately, as attractive as such an answer might appear, a naturalistically informed account of human knowledge recognizes that human knowledge serves human interests.[8] It further recognizes that there are different types of human interests which generate different types of human knowledge. This point was earlier introduced in terms of the four modalities. Unfortunately, to adequately develop this issue would require a lengthy paper in its own right. What is important in the present context is that knowledge of the empirical world tends to be governed by pragmatic/instrumental interests while our theoretical concern with reality *per se* tend to be governed by our interest in establishing a shared communal worldview. Whereas in the first case – instrumental interest – quantification becomes a means for grasping the objects of experience, in the second case quantification becomes a means for establishing consensus. Here I must ask for the readers indulgence in an attempt to support this last point.

Earlier it was argued that the empirical world lends itself to quantification because it presents us with a world of potential equivalencies. It allows us to count because it presents us with discrete objects rather than a flow of intermingled experiences. Whether these empirically experienced objects are real or not, they are clearly out there. (We sometimes get into arguments as to whether our experiences are 'real' or imagined, but the more classical issue is whether we experience reality or false shadows. Such shadows may be false, but they are out there.) Counting, consequently, is a process which entails distancing ourselves from that which is counted. Given that personal bias may serve to undermine consensus, quantification in and of itself conveys the impression that that which is counted is somehow independent of personal/subjective bias. Whereas the process of quantifying conveys the sense of 'objectivity' in a derivative manner, i.e., by being commonly associated with experience of objects, it also conveys the sense of 'non-subjective' insofar as it entails a distancing of that which is counted from the agent that counts. It may be argued here that I am attempting to split hairs, but the distinction made is not without significance. It explains, for example, how numbers and mathematics could be seen as entailing 'reality' quite separately from the empirical world. It is as if the process of quantification acquires in its own right the sense of independence and separateness originally inherent in our experience of spatial-temporal objects.

While the thoughts just expressed may begin to sound somewhat mystical, the situation I am trying to describe is really quite straightforward. Counting is a form of naming and naming is a process whereby inherent powers and/or dispositional tendencies are attributed to some other (other than self) thing. Counting like naming endows that counted and/or named with an existence independent of one's own experiences of the thing in question.

Here it may legitimately be asked why we should put value on such independence. Why need the 'real' be independent of our experience of it? In point of fact it need not. The key point here is to recognize that in our pursuit of knowledge our primary concern is not to obtain an 'accurate' account, but rather to obtain an account upon which we can agree. The more we can set limits on personal idiosyncratic imputs the greater the possibility for communal consensus. Here we come across a peculiar quirk in the history of science. Scientific accounts carried the day against earlier theological accounts not because they told us more about 'reality', but because they offered accounts which lent themselves better to consensual agreement. The fact of the matter is that the 'what' which the new science presented was, at that time, generally not accepted as 'reality' but rather merely as 'appearances.' It is we who have accepted the world of Kantian appearances as reality. We have accepted it as reality because it provides the focus around which shared accounts can be generated. One of the reasons it lends itself to such shared accounts is that it lends itself readily to quantification.

To say that the world of objects lends itself to quantification may appear superfluous in light of our earlier discussion. I would argue, however, that although quantification can be seen as initially grounded in experiences of spatial-temporal objects, it possesses its own heuristic and explicative power insofar as it reifies through distancing from subject that which is quantified. I would further argue that it was, and to a large extent still is, the process of quantification *per se* which conveys the sense of 'reality' rather than our modern empiricism. The success of modern empiricism was that, not surprisingly, it lent itself to quantification. It was a case of quantification coming home.

That quantification can and should be seen as a means of grasping reality quite apart from empiricistic methods despite their developmental relationship is evidenced by the numerous non-spatial-temporal contexts in which it is used; a number of such contexts were noted earlier. I would like to examine some of these examples in more detail. More specifically, I would like to examine quantification in the context of biblical scholarship, performance rankings, and auctions. (There is no logical relationship among the three; they merely represent some areas of my own interest which relate to the issue under discussion.)

Few objects have been quantified to the extent to which the five books of Moses have been quantified. There is, in fact, a special name for the process, namely, gematria. Gematria takes advantage of the fact that in Hebrew each letter of the alphabet can be given a numerical value. This allows for all sorts of new equivalencies, i.e., a specific phrase can be shown to have the same numerical value of another phrase. It is alo possible to assign other quantities to items based upon the number of words in a section, or where in a section an item appears and seek out other equivalencies. Since there is a vast literature on the subject, I can do no more than indicate the most common procedures used. In reality the creative imagination which has gone into this type of scholarship is awe inspiring.

In structure, gematria is similar to other types of mystical/magical uses of quantification. In nearly all such cases I would suggest that it is possible to distinguish three interrelated objectives. First, by assigning a number to a verse or some textual item, that item is given a status independent of its original context; it is no longer merely part of a particular story but a type of entity in its own right. It is, for example, an eighteen. Secondly, this independent status serves to free the item not only from its original context, but to give it an independence *vis-à-vis* subjects. The non-quantitative meaning of the item may be such that it generates a good deal of 'subjective' interpretation, but such 'subjective' input can be limited if not eliminated by focusing upon the item's quantitative meaning. Thirdly, by acquiring such independence from both its original context and subjects, such items can be related in new ways to other quantified items. While we often assume such characteristics to be associated with spatial-temporal objects, it is important to recognize the extent to which these characteristics can be derived directly from number *per se*.

While anyone who has looked into gematria even superficially is likely to be impressed by the subtlety of the process, they are also likely to wonder about the utility of the process. Put quite bluntly, to many, including a good number of biblical scholars and committed believers, the whole process is sheer nonsense. Even such critics, however, generally recognize the seductiveness of gematria. Here we come to another crucial aspect of the quantification process. Given that the 'objectivity' of such quantities as those involved in gematria – and I might add other forms of mythical/magical analysis – is imposed upon the data, I would argue that we are not here dealing with a process which claims to be getting closer to what is 'empirically' out there.

Insofar as the process presents itself as a mean for grasping reality, the reality clearly entails some form of transcendence. (I think that by and large this is exactly what is going on, but it is an issue I don't want to get into here.) What I think is important about this example is that while there may be some element of trying to grasp what is out there, in a non-empirical sense, what appears to be even more important is the desire to establish some sort of explanatory account which lends itself to specified procedures and consequently to a broader consensus. That is, I would claim that such examples underscore the point made earlier that quantification serves as a means for establishing agreement rather than as a technique for accounting for empirical reality.

The extent to which quantification functions as a means for establishing agreement is supported by the second set of examples I want to briefly examine, namely the use of quantification in ranking performances. Here again there are many instances of the process from a wide range of sporting events to I.Q. tests to rankings in casual conversation. The process is familiar; what is in question is the purpose of the process.

In some cases, such as I.Q. tests and other forms of 'objective' exams, it may be argued that the purpose is to measure accurately some capacity. By and large, however, I think it is fair to say that most experts in the field are willing to admit that I.Q. tests measure how well you do on I.Q. tests though such capacities may also be correlated with other performance skills. The primary purpose of such instruments is to create a means for comparing otherwise heterogeneous capacities and/or performances. This is clearly what is happening when a particular number is assigned to a complex dance, gymnastic, or skating routine. This is not to deny that some attempt is made to encorporate within the scoring process means for grasping what is claimed to be measured; the most important objective, however, is to assign a quantity. This act not only allows the various judgments to be integrated by, for example, working out the average score, but also conveys the impression that we are now confronted with a level of performance rather than a judgment about that performance. There may very well be disagreement regarding the score assigned, but better such disagreement than no score.

While the examples given allow for a wide range of interpretation, I think the single most important point they suggest is that quantification's primary objective is the establishment of a shared account rather than a more empirically accurate account. The goal is not primarily to classify our experiences in a theoretically sound manner, but rather to render them publically accessible The final judgment may be experienced finally as inaccurate, but it will be a shared judgment.

Nowhere is the relationship between the desire for a consensual judgment and quantification more clearly revealed than in auctions. Here we have a clear example of consensus winning out over an assumed 'objectivity.' It may be argued that I am here presenting a false dichotomy; the objective value of the thing is what is bid for it. While there is clearly a sense in which this is true, there is also clearly a sense in which it is not true. Anyone who has spent any time at auctions knows that there is usually a clear sense among those present as to whether the item was sold cheaply, about right, or overpriced. The more fungible the item the more likely that there will be a sense of 'fair' price, but even when dealing with exotica and one of a kind items one can normally 'feel' whether the price was too high or too low.

Given that it is generally possible to assess the approximate value of an item independently of the auction process, the question arises why bother with the auction at all. The reason is two-fold. First, while most participants may be willing to admit that the item in question may be said to have an approximate 'true' value, the actual process of determining that value may be very cumbersome and time consuming. Second, though it may be possible to establish a broad consensus, what is called for is a consensus between buyer and seller which need not be the same consensus as the general consensus. It should be noted, however, that this situation is not unique to auctions. It may be said to hold in any economic exchange where bargaining goes on.

What is peculiar to the auction situation is that the exchange process becomes the means for establishing the value of the item rather than the establishment of a price being the means for enabling the exchange. It is this difference which makes auctions of common interest whereas most instances of bargaining are of little interest to anyone but those directly involved. We are all interested in material goods and who owns what, but such interests pale when compared to our interest in consensual meaning given to the world within which we live. If G. H. Mead, Levi-Strauss, and Habermas are correct – and I think they are – that the primary objective of the species is the creation and maintenance of shared symbolic universes, then we can understand the fascination with auctions since this is exactly what is going on in any auction.

While the various examples given above differ from each other in a number of ways, they are similar insofar as they clearly reveal that the process of quantifying phenomenon need not entail any attempt to more accurately represent the phenomenon in question. Put slightly differently, it should be clear that the commonly claimed scientific reason for imposing quantitative values upon experiences does not apply to these cases. I would further suggest, however, that the 'scientific' claim of greater representational accuracy is often not the governing objective even in so-called scientific contexts. 'Scientific positivism' has clearly put great emphasis upon quantification; moreover, it has been quite successful in carrying out this objective. Do such accounts, however, better explain what is happening or do they merely lend themselves better to consensual agreement?

What makes scientific positivism – the notion that reality is the empirically experienced spatial temporal world – so attractive is that it lends itself naturally to quantification. The power of quantification, in turn, is that it is a form of labeling which lends itself to consensus. This is not to deny that quantification is both a natural and useful technique in the attempt to symbolically represent the world of spatial-temporal objects. This point was acknowledged earlier. The key point here is that the essential attractiveness, one might say seductiveness, of 'scientific' quantification, especially when applied to other types of phenomenon rest with its consensual powers rather than its representational powers. This is admittedly a bold claim which is not easily defended. I might simply note, however, that the movement toward ever more quantification appears to be positively correlated with the emergence of distinctive 'schools' of thought where the need for consensual agreement becomes ever more important.

Though consensus is seldom the stated goal of a scientific enterprise, it should be noted that there is nothing inherently unsound about such an objective. A strong case can be made for the position, in fact, that a broadly based consensus is eminently better suited as the criteria for 'truth' than some form of dubious representationalism. What is particularly troubling about the use of quantification in the social sciences is that it unnecessarily carries with it the naive empiricism commonly associated with quantification in our fairly inaccurate conception of the physical sciences. In short, where quantification is used to construct explanatory accounts as in the case of certain forms of mathematical modeling, it may be the case that we are only playing games, but generally no misrepresentation results. When, however, experiences are reduced to quantitative categories and inappropriate distancing occurs, misrepresentation is often the result.

As noted earlier, the ever greater tendency to quantify experience is not limited to the physical or social sciences; it has become ingrained in our everyday lifes. As a corollary of this, we also tend to impose spatial-temporal/object definitions upon everyday experiences. The two tendencies have become so interrelated, in fact, that it is often a case of what came first, the chicken or the egg. Whatever the answer, I would suggest that it is clearly the empiricistic tendency which is most lamentable, since it leads not merely to possible misinterpretations, but to misrepresentations of reality. I would suggest that the same situation holds for the social sciences.

Numbers are specific types of names and convey specific meanings, often highly qualitative in character; as such they are powerful tools in generating explanatory accounts. This holds true whether or not the form of the account is itself mathematical.[9] There are also times when social scientists' focus of concern is most properly defined in terms of spatial-temporal/object categories and hence lends itself to descriptive quantification; certain types of demographic studies would be such an example. When, however, these conditions do not hold and when we see a number of practitioneers attempting to quantify materials which do not lend themselves to such practices, it is more likely than not that we are not observing an attempt to grasp reality, but rather an attempt to reach a group consensus.

NOTES

[1] The essays in this volume I believe sufficiently attest to this fact.

[2] In addition to Aaron Cicourel's *Method and Measurement in Sociology* (New York: Free Press, 1964) see collection by Daniel Lerner and Harold Lasswell *The Policy Sciences* (Stanford: Stanford University Press, 1959).

[3] See Roy Bhaskar's *A Realist Theory of Science* (Leeds: Harvester, 1975) and *The Possibility of Naturalism* (Brighton: Harvester, 1979).

[4] See 'On the Sociology of Mind' by Charles W. Smith in Paul Secord's *Explaining Human Behavior* (Beverly Hills: Sage Publication, 1982).

[5] George Herbert Mead's *Mind, Self and Society* (Chicago: University of Chicago Press, 1934) remains the classic formulation of this position. For a more recent formulation see the Introduction of Anthony Giddens' *New Rules of Sociological Method* (New York: Basic Books, 1976).

[6] Immanuel Kant, *Critique of Pure Reason*, trans. Norman Kemp Smith (New York: Macmillan, 1958).

[7] Piaget touches on this issue in a number of different works. See particularly *The Child's Conception of the World* (Paterson, N.J.: Littlefield, Adams, 1960).

[8] This is another issue which has been dealt with by many persons. See particularly Jurgen Habermas, *Knowledge and Human Interest* (Boston: Beacon Press, 1972).

[9] For what is still the best non-mathematical discussion of quantification see Georg Simmel, *The Sociology of Georg Simmel*, ed. and trans. by Kurt H. Wolff (New York: Free Press, 1950, 1964) pp. 87–177.

CHARLES W. LIDZ

'OBJECTIVITY' AND RAPPORT

INTRODUCTION

Although quality field research has gained a growing place in sociology in the last decade, it faces a continuous challenge as to whether or not it is really 'scientific.' Given the enormous prestige that the quantitatively based physical sciences have in our society, this is hardly surprising. Thus, in spite of the many advantages of techniques of research that allow access to complex structures of meaning, qualitative research faces a constant struggle to maintain its legitimacy with both consumers of social science research and each new generation of students.

While this battle is being fought with those who do not see the value of qualitative methods, there has been a substantial internal conflict about the direction in which qualitative methodologies should develop. One group of sociologists has argued that the proper use of these methods depends primarily on the interpersonal abilities and theoretical training of the researcher. For this group, training consists of fieldwork experiences with supervision from more experienced fieldworkers. Classroom training focuses on theory, not method (e.g., Emerson, 1983). On the other hand, there is a substantial group of qualitative sociologists who are not so comfortable with leaving behind such 'scientific' concepts as reliability and validity. They have tried to tread a fine line by remaining committed both to understanding the meaning of the events under study to the participants and, at the same time, improving the scientific status of their fieldwork procedures. While there has been relatively little explicit defense of this difficult position, the continuous pressure in that direction is visible in such diverse areas as the growing use of computerized analysis of field notes (Seidel and Clark, 1984), the use of video and audiotapes as a means of recording data, the development of organized teams of observers (Douglas, 1976), and the use of statistical analyses in analyzing field research data (Becker, 1958).

While critiques of qualitative methodologies from the latter perspective are often received with the same enthusiasm that greeted left-wing critiques of American society during the Cold War, we will try to show here that there are sometimes good reasons to consider some of the standard scientific questions as applied to field research. This paper focuses on the classic positivist

43

Barry Glassner and Jonathan D. Moreno (eds.),
The Qualitative-Quantitative Distinction in the Social Sciences. 43–56.
© 1989 *by Kluwer Academic Publishers.*

objection that fieldworkers cannot be 'objective' in their reports of their findings. Rather than simply restating that 'objectivity' is not a simply achieved goal for quantitative sociologists either, we will look at some of the sources of 'bias' that might enter into field reports and their analysis and consider how they may best be managed.

The idea of objectivity is a conceptually peculiar one. How is it possible for a subject to be objective? Are we not inherently bound to our subjectivity, and thus is not objectivity a delusion? Weber (1949) provided what remains the most adequate answer by suggesting that objectivity involves a commitment to the universalistic value commitments of the scholarly community in the context of which the research is done. In this context, gathering objective data and analyzing it objectively means to gather and analyze data in the context of the universalistic methodological and meta-methodological commitments of the sociological community.[1]

This paper raises the question of whether or not the fieldworker's traditional effort to build a close rapport with those under study is a fruitful way to gather 'objective data' and is consistent with 'objective analysis' in the sense just discussed. In doing so, it will focus on the concepts of gift, reciprocity, and solidarity.

RECIPROCAL RELATIONS

The work of Emile Durkheim and his pupil Marcel Mauss provides the basis of a theory of reciprocal relationships, how they come about, and the constraints that they impose on each party. One focal point of this theory is Mauss' essay *The Gift*. In it Mauss showed that 'although the prestations and counterprestations take place under a voluntary guise, they are in essence strictly obligatory.' (Mauss, 1954:3). Thus the giving of anything of social value, including, as Mauss noted, courtesies, rituals, entertainments, promises, and favors of all sorts, obligates the one to whom the socially valuable gift is given to return a gift of equal value.

Equally important for our purposes, there exists a more diffuse bond that is created by such exchanges. Both in Durkheim's analysis of the non-contractual elements of contract (Durkheim, 1933) and in Mauss' analyses of the gift exchange, it is clear that exchanges not only involve a short-term obligation to return the exchange but a diffuse bond, which Durkheim referred to as 'solidarity,' between the two parties. Simmel, speaking of the 'gratitude' that develops between two parties to an exchange relationship, describes the phenomenon as well as either Durkheim or Mauss:

The irredeemable nature of gratitude shows it as a bond between (people) which is as subtle as it is firm. Every human relationship of any duration produces a thousand occasions for it and even the most ephemeral ones do not allow their increment to be lost ... the sum of these increments produces an atmosphere of generalized obligation (Simmel, 1950:395).

Thus we can see that the giving of gifts of whatever sort binds the participants in the relationship in two ways. First it results in the specific obligation that the gift must, in some way, be returned. Second it binds with a long-term solidarity that obligates one participant to the other in a diffuse way. These obligations are not casual ones. How a person lives up to such obligations is closely connected to the individual's publicly known moral character. A respectable member of the community is expected to fulfill these obligations unless good reasons can be given why their violation is unavoidable or otherwise justifiable.

This paper considers what types of generalized obligations a fieldworker undertakes to the people being studied and what the consequences of these obligations are for the generation of 'objective' data from fieldwork.

FIELDWORKER AS GIFT RECIPIENT

This section considers the 'gifts' the fieldworker typically receives in ethnographic research and the obligations that they impose on the fieldworker. In doing so, it will draw both on the literature in the area and on the author's experiences doing research in the criminal justice system, in psychiatric facilities, and medical hospitals. Subsequently, we will be able to discuss the ways in which the fieldworker may reciprocate those gifts.

One way of understanding the importance of a gift is to consider the needs of the gift recipient for the gift. Thus it is useful to begin by discussing the difficulties in the role of fieldworker. Probably nothing is more often the topic of legend and lore among field researchers than their experiences gaining entree and acceptance among their respective subject populations. In comparison to, for example, Mayberry-Lewis' (1965) experiences almost starving to death among the Shavante Indians in Brazil because they had no interest in teaching him how to find food, let alone providing it for him, most sociologists' difficulties are minor. Nonetheless, entering the field is an anxiety producing event in any setting. Even in a rather mundane setting like a courtroom or a psychiatric admission service, the social difficulties are quite substantial. For example, can one interrupt a conversation between two police officers to ask what they are talking about, or is the particular conversation so important that one should wait until later? Can the observer enter the judge's chambers or is that too private a setting? How about listening in on the psychiatrist's telephone call with the parents of a patient? When studying a probation officer's contacts with his clients, where does one sit in a room that does not have an extra chair? Which statements are meant to be funny? ironic? sarcastic? Anyone who has ever done fieldwork knows how embarrassing it can be not to know how to deal with these issues. Because of this sort of difficulty orienting oneself socially in a fieldwork situation, it is not unusual for a fieldworker to end up playing the role of pet or clown in the presence of his or her subjects (Daniels, 1972; Bosk, 1979).

It is equally difficult to orient oneself cognitively to what is going on. The classic cultural anthropologist who has to learn an unknown language as well as all of the particulars of a tribal culture has the greatest difficulties in this area. Many contemporary fieldworkers have learned how alien and lost one can feel in someone else's turf ten blocks from one's academic office. While Bar Hillel (1954) may be right that all expressions are inherently indexical, my own experience is that no one's talk is quite as indexically particular as two narcotics detectives exchanging street gossip. Even after three weeks observing the same officers, I often found myself having no idea what they were talking about for minutes on end. In any setting there are references to places, events, and background knowledge that the observer does not know. Conversations take their meaning from particular projects and situations that the observer often cannot understand unless they are explained in more detail than the participants would otherwise find necessary. For example, clinicians in the admission unit of the psychiatric hospital that I studied routinely speak of '302ing' someone. This particular verb turns out to derive from the section of the state regulations that permits emergency involuntary commitments.

The problem of gaining entree to the setting, getting the confidence of the participants, and, most importantly, gaining a basic orientation to the subculture are usually solved in part by looking for a person to serve as an 'informant.' In spite of the name, this role is only partially a cognitive one. The person who serves as primary informant may also function as 'gatekeeper' and sponsor the ethnographer with friends and acquaintances by testifying to his or her good character as well as seeing to it that the ethnographer does not make too many embarrassing mistakes. Elijah Anderson has described this phenomenon as well as anyone when describing the role that his primary informant, Herman, played in facilitating his entrance into a group of men who hung out outside a liquor store on the South Side of Chicago. His major breakthrough began when Herman told the group that he had invited Eli to attend a party at his place of work:

> The men of the group continued to check us out. Herman was treating me as a friend, as an insider, as though my status in the group were somehow already assured. Certainly Herman would not invite just anyone to a party at work ... After this demonstration by Herman I could feel the others in the group warm up to me; they looked at me directly and seemed to laugh more easily.
>
> I began to feel comfortable enough to stand around and listen to the banter, but not to participate in it, my reticence reflecting my sense of visitor status. But I did laugh and talk with the fellows, trying to get to know people I had often wondered about during the past four or five weeks. They were people I had seen around but never attempted to 'be with.' ... When Herman introduced me to his friends I was in effect being sponsored and in many ways this made my status passage into the group relatively easy ... When Herman sponsored me and invited me to the party, I assumed some usefulness for and responsibility toward him (Anderson, 1978:16–17).

In short, the good informant not only has to provide good information but has to spend considerable time and effort on helping orient the ethnographer and to risk his or her reputation within the community. Obviously this often grows into a substantial role.

THE ETHNOGRAPHER'S OBLIGATIONS

Viewed as a gift, these actions of the informant can hardly be treated as negligible presents. As a good 'host' in a strange setting, the informant performs tasks that allow the observer a certain amount of comfort in the setting and preserve the observer's face and claim to an acceptable status in a situation in which the observer will spend a lot of time.

It is hard for a researcher not to have a deep sense of debt to the informant when leaving the field. It is, of course, true, as Rosalie Wax (1952) has noted, that the person being observed often finds the role gratifying, and Anderson (1978) has noted that the ethnographer may present the informant with an opportunity to elevate his or her status *vis-à-vis* friends and community. Even when observing someone with a higher social status, as I did when studying surgeons, the observer's presence may provide the subject with an opportunity to brag about having 'my own sociologist.'

None of this, however, does much to mitigate the sense of obligation. Gift giving is an active process, and the ethnographer must find an active way of returning the gift.

How can the observer deal with his or her sense of obligation to the informant? As Mauss noted, the standard way of dealing with this sort of obligation is to return the favor in kind. But the sociologist or anthropologist can hardly suggest to a shoe salesman, prostitute, or Shavante tribesman 'Why don't you stop by and observe me sometime.' Instead, we typically do something that is similar to what Emily Post suggested as a way of returning a favor when you are not in a position to reciprocate directly.

Let us say that the Greathouses invite their friends the Onerooms to dinner, then to lunch, then to dinner again, throughout every month of the year ... Mary Oneroom (must put) herself out to do a dozen kindnesses of one sort or another ... This 'payment of oneself,' as the French say, is an important phase of hostess-and-guest obligation ... (Post, 1945:432).

The absence of a clear-cut procedure for repaying the obligation that is owed means that Ms. Oneroom and her husband must take on the role of the 'gratefully indebted' as part of the repayment scheme. This involves continually being on the lookout for various ways of making up an obligation which, because of differences in position in the relevant community, can rarely be entirely repaid. One of the easiest ways in which this can be done is by ego reinforcement or, to put it more bluntly, flattery. The fieldworker is in precisely the same situation with the informant. The fieldworker is never in a position to reciprocate all of the favors owed in a direct way to the informant and thus must take on a 'gratefully indebted' role and spends a lot of time flattering the informant.

As noted earlier, there are two levels of bond between the two parties to an exchange. The obligation to reciprocate the giving of gifts is only one of them. The second level concerns a basic solidarity that is generated in any exchange. This involves the development of a diffuse sense of trust that the other actor involved in the exchange shares a basic orientation toward the project that the two parties undertake together. This is very similar to what ethnographers call *rapport*. Ethnographic methodological treatises have long recognized the existence of this bond and have encouraged the fieldworker to develop it as much as possible. For example, a recent text on field research suggests:

(It is) essential for researchers to win the trust and confidence of their subjects; that is, to establish rapport with them ... In the ideal situation participant observers develop close and honest relationships with all of the participants in the setting in which they conduct research (Bogdan and Taylor, 1975:46, 48).

Similarly, Johnson reports that in studying welfare workers, he routinely encouraged his subjects to believe he shared their values.

In the context of informal conversation, usually when the social workers and I were out in the field together, I would communicate the general idea that I was a person with a sympathetic understanding of social welfare. I often said that my previous experiences had lead me to understand that many things do not work in the fashion presented by the official lines in the textbook (Johnson, 1975:101).

In my own work I have found myself expressing disdain for 'bureaucratic rules' that I had previously supported, making cynical comments about Constitutionally required courtroom procedures and discounting the value of the services provided by organizations that were paying my salary. All of these were designed to demonstrate to someone I was studying that I shared their perspective on a particular feature of their environment. Most fieldworkers seem to do this in order to demonstrate to those they are studying that they can be trusted with organizational or in-group secrets.

What has not generally been appreciated about rapport is that solidarity and exchange are both reciprocal relationships. They involve obligations for both sides. We have discussed the important 'gifts' that the informant provides the ethnographer. What should be apparent from the Johnson quote is that the observer also commits him or herself to certain obligations. *The existence of rapport not only means that the informant is committed to open up a private world to the observer but also that the observer is diffusely committed to view that world in a sympathetic light.* Let us be very clear about that commitment. It is a commitment to view that world *particularistically.* This commitment stands in stark contrast to the commitment to objectivity, that is, to see the information gathered through the universalistic perspective of sociology. Like other such commitments, we are often only half aware of it but it frequently has a profound effect.

This particularistic commitment involves three problematic elements for the fieldworker. First, viewing the information particularistically involves not directly challenging the factual claims of the informant. At the very least, this restricts substantially the types of questions that one can ask and thus limits the data that can comfortably be collected. Second, it involves the assumption that the observer accepts the same value commitments that the informant does. This becomes particularly problematic in the analysis when the ethnographer is involved in policy research that is committed to different values, e.g., a study of welfare case work that is committed to improving compliance with organization rules. Finally, it obligates the fieldworker to use the knowledge gained from the project in a particularistic manner. At the very minimum this means that nothing that the observer does with the findings should damage the interests of the informant. More generally it means that the analyst feels obliged to present the informant the way he or she would present a friend. downplaying 'faults' and 'weaknesses' and highlighting 'virtues.' The fieldworker can thus be said to be obligated to the person or persons whom s/he observes both because of the difficulty of reciprocating specific favors and because of the generalized bond of rapport that involves diffuse particularistic obligations. *Both of these obligations conflict with the sociologist's universalistic obligations to the research enterprise to treat all observations as 'data.'* How are these conflicts resolved?

GOOD FAITH AND BAD FAITH ANALYSES

Although it is not always an easy practical task, there is no reason in principle why a fieldworker should not be able to maintain at least the appearance of upholding these conflicting commitments while doing fieldwork – albeit not without some restrictions on the sorts of data that can be collected. However, when the fieldworker turns analyst, the problem of conflicting commitments becomes more difficult.

There are two ways in which we deal with the obligations that we have undertaken in the field, which I will call 'good faith' and 'bad faith.' In spite of those terms, I do not mean to imply any value judgments. In situations in which one has conflicting commitment, like this one, it is often true that good faith in dealing with one set of commitments implies bad faith in dealing with another.

The 'good faith' alternative involves living up to the obligations to the people who facilitated the fieldwork – the informants, gate keepers, and the other subjects of the research. However, it also involves giving up the professional norms of unbiased analysis. There have been many efforts to deal with this conflict. For example, Howard Becker (1966) argued that sociologists of deviance should function primarily as advocates for the populations that they study rather than pretending to be doing disinterested science. In a somewhat similar vein, Gould *et al.* (1974), having observed drug users and both psychiatric and legal control agents, presented their findings by describing, in the same volume, separate and conflicting accounts of the drug problem from each perspective. However, to the degree that we are pursuing a 'scientific' sociology, neither of these procedures is a totally adequate solution to the problem.

Another frequently used solution to the conflict might be called a 'retreat into academic irrelevance.' Using this approach, one selects a narrow, scientifically interesting but practically uninteresting question and analyzes the data only with reference to that question. Assuming for the moment that one really is uninterested in the policy or practical issues raised by one's research, this might seem to solve the problem. But while the difficulties with this stance are subtle, they are nonetheless real. One frequently finds oneself playing elaborate word games to hide what one is talking about. For example, in my own work studying the ways in which narcotics police go about the management of heroin use, I ended up focusing on the non-legal emergent rules of interaction that characterize the relationship between the detectives and addicts. Whatever the merits of this as a scientific issue, it seemed to neatly allow me to avoid such difficult issues as violence, illegal arrests and searches, perjury, racism, planting drugs on suspects, and other ordinary but controversial vice squad behaviors. However, it is hard not to discuss such issues when talking about the balance of power between police and addicts that underlies the tacit rules of interaction. Nonetheless, my analysis made only the most cursory and passing mention of any such behaviors (Lidz, 1974). In retrospect it is hard not to think that 'good faith' to my informants interfered with the practical utility, if not the scientific validity, of the analysis.

If one ignores the ethical problems that it raises, the 'bad faith' alternative seems, on the surface, to be the better solution. This alternative involves fidelity to the norms of scientific objectivity and treating one's data in a universalistic manner. The obligations one undertook while doing fieldwork are dismissed as not relevant either because one's subjects will never read one's 'scientific' papers or because the commitments were never 'real' in the first place but done solely as a technical device to comply with a technical requirement of such research called 'getting rapport.' Besides, one warned one's subjects that one would analyze the data scientifically (cf. Douglas, 1976).

Yet detaching oneself from the obligations one undertook during fieldwork seems to be easier in theory than in practice. As noted earlier, one of the ways that we evaluate both ourselves and others concerns the ways in which we carry out our obligations to others. Unless we can find reasons not to carry out obligations, we must judge ourselves and be judged by others as persons of dubious moral character. Appeals to scientific norms are only a partial excuse because, of course, the fieldworker knew about such norms when he or she led the subjects to believe that they would be dealt with particularistically. A better excuse would be for the fieldworker to show that the subjects are not deserving of such treatment. Thus the secrets and the 'muck' that are discovered through carefully developed rapport become the basis for the justification of the betrayal of the observer's end of the rapport. The resulting analysis becomes distorted by the moral need to overemphasize and decontextualize the 'muck' that was discovered. Perhaps equally serious is that the breaking of obligations leads to an alienation of the ethnographer from the informant. Thus the ethnographer finds it increasingly difficult to empathize with the values and motives of those individuals who were the subjects of the research – precisely what many ethnographic studies are designed to assure. The frequency with which ethnographic studies end up as either muckraking or apologies reflects the difficulties of managing the role conflicts built into the role of the ethnographer.

MANAGING A ROLE CONFLICT

In one of the finest books on the task of doing fieldwork, Johnson (1975:84) wrote that 'There is widespread consensus among field researchers that rapport or trust between the observer and the members is an essential ingredient for the production of valid, objective observations.' Although Miller (1952) observes that what he calls 'overrapport' can prevent the gathering of some sorts of data and Roadburg (1980) observes that close personal relationships can make terminating the fieldwork difficult for both the researcher and the subject and Schwartz and Schwartz (1955) warned against the bias of overinvolvement with the subject, Johnson is correct that the consensus is almost universal. However, if the above analysis of rapport is correct, the closer the rapport between the ethnographer and the subject, the less likely that the eventual analysis will be 'objective.' To put it another way, the greater the fieldworker's obligation to the informant or subject, the greater the role conflict with his or her commitment to scholarly or professional norms. If one's goals are not to present an objective version of what one is studying, this may not be a problem, but for those who seek such an account, the question is what might be done to minimize the damage.

We might begin by recognizing that whatever the personal pleasures of developing close relationships with those whom we are studying, it is not necessarily a methodological necessity. How much rapport is necessary depends on what one is studying, what sort of information is necessary, and the specifics of the setting. For example, Zerubavel's study (1979) of the temporal structure of a hospital required permission from the hospital administration to do the study and, since he had to hang around the nursing station, the acquiescence of the nursing staff. In general, however, his research did not suffer in any way from his maintaining a substantial distance from the people whose behavior he was studying. The sort of information that he needed from the hospital staff, e.g., how shift schedules are made up, required little intimacy with the staff. Likewise, in spite of the fact that it involved observing much illegal behavior, my own field research on plea bargaining only required permission from the chief prosecutor, who seemed to care little about what I did. I sat and watched negotiations between the assistant prosecutors and defense attorneys but was involved in very little interaction with either group beyond the usual interpersonal pleasantries.

Obviously some studies require more rapport than others, but the assumption that valid data can *only* be collected in the context of an intimate relationship seems overstated. An important feature in both of the above studies was that they involved the study of busy people in one physical location. If one could get permission to sit in one place, one could do the study. Another case in which one can do studies without extensive rapport is one that involves almost no behavior that the subjects see as in any way controversial. For example, a current study involves trying to determine how psychiatric clinicians determine what is relevant and irrelevant in what a patient tells them. Asking questions about that seems to the clinician to be nothing but a series of mundane questons about ordinary events and seems to require only a minimal working relationship. Many sociologically interesting questions can be studied without extensive rapport.

Another way in which one can study some processes without the usual difficulties of entree is to do the research in a setting that routinely is invaded by students who are being socialized into the enterprise one wants to study. While such a setting can obviously not be taken as synonymous with a non-teaching setting, it seems reasonable to assume, for example, that medical examinations are not radically different in structure when done in the presence of students. Students also provide the fieldworker with someone else to ask the simple-minded questions.

Another partial solution to this problem is to segregate the role of fieldworker from the role of analyst. In my current research, we use a team of fieldworkers whose job it is to take shorthand notes which then provide something like a transcript of the interaction between those being observed. They also record generalized background observations about the interaction. The analysis is performed by different members of the team. While this is obviously not a universal solution to the problem, and it raises other problems, it is a useful way of separating the analysis from the emotional commitments of the fieldworker in some circumstances.

A more traditional solution to the problem of the loss of objectivity is for the fieldworker/analyst to be supervised by another professional. This is, of course, most prevalent in the relationship between student and dissertation supervisor.

Finally, as has often been noticed, there is a parallel between the relationship between the fieldworker and the informant and the psychotherapist and the client. In both cases a minimally involved professional becomes acquainted with intimate details of an individual's life, and the relationship may become an emotionally close one. Psychoanalysts, who pride themselves on the detachment with which they can analyze the information provided by the patient, have treated the ability to analyse the 'counter-transference' as an essential skill in which they are trained. While this is probably less central to our task, it is probably true that a conscious awareness of the problem would substantially reduce the problems that rapport might cause for the objectivity of the analysis.

CONCLUSION

The ethnographic concern with good rapport may arise less out of scientific methodological needs than from other sources. First, we often share with the rest of the society the assumption that what is really interesting are the in-group secrets that we can gather about any group that we are studying. We find it interesting to know about the ways in which tradesmen get around paying taxes, policemen violate the law, doctors ignore their patients' desires, and friends gossip about the sins of their neighbors. While this interesting behavior requires close rapport to observe, it is amply documented in the literature and is not inherently more interesting for the development of an adequate sociological understanding of the world than any other behavior.

A second reason that we are over interested in rapport may have to do with a certain type of 'macho' ethos among fieldworkers. We like to feel that, unlike our quantitative colleagues, we can take care of ourselves outside of the groves of academia. We take our close relationship with burglars, nudists, and company presidents as a sign of our ability to function outside of academia in the 'real' world. Finally, inasmuch as having close rapport with a subject comes down to being liked, it reflects a simple affiliative need that most people have. All of the above are good reasons for close rapport, but they are not scientifically essential ones.

It is important to be clear about what I am not saying here. I do not mean to imply that it is necessarily a bad thing to develop a close relationship with the people one is studying. If one seeks to understand the personal or private features of the individual's life or to study work situations which cannot be publicly known without causing substantial public embarrassment, it is important to gain the trust of one's subjects. However, rapport has its costs. I have suggested that there is some reason to believe that our emphasis on rapport with those we study has led to substantial difficulties in general ethnographies. Be that as it may, ethnographers would be well advised to consider the question of what sort of relationship they want to develop with their informants before they begin the study and to weigh the risks of overrapport with the advantages that are to be gained.

ACKNOWLEDGEMENT

I am indebted to Mary Carter, Joel Frader, Carl Johnson, John Marx, Dan Regan, and Eviatar Zerubavel for comments on earlier versions of this paper.

NOTE

[1] See also Talcott Parsons' excellent rendering of Weber's position on this subject (Parsons, 1965).

REFERENCES

Anderson, Elijah, *A Place on the Corner*. Chicago, University of Chicago Press, 1978.
Bar-Hillel, Yehoshua, 'Indexical Expressions,' *Mind* 63, 359–79, 1954.
Becker, Howard, 'Problems of Inference and Proof in Participant Observation,' *Amer. Soc. Rev.* 23(6), 652–660, 1958.
Becker, Howard, 'Whose Side Are We On,' Presidential Address to the Society for the Study of Social Problems, 1966.
Bogdan, Robert and Steven Taylor, *Introduction to Qualitative Research Methods*. New York, Wiley, 1975.
Bosk, Charles, *Forgive and Remember*. Chicago, University of Chicago Press, 1979.
Daniels, Arlene K., 'Military Psychiatry: The Emergence of a Sub-Specialty,' in *Medical Men and Their Work*, edited by Eliot Friedson and Judith Lorber. Chicago, Aldine-Atherton, 1972.
Douglas, Jack, *Investigative Field Research*. Beverly Hills, Sage, 1976.
Durkheim, Emile, *The Division of Labor in Society*. Glencoe, IL., Free Press, 1933.
Fielder, Judith, *Field Research*. San Francisco, Jossey-Bass, 1978.
Gould, Leroy C., Andrew Walker, Lansing Crane, and Charles W. Lidz, *Connections: Notes from the Heroin World*. New Haven, CT, Yale University Press, 1974.

Johnson, John, *Doing Field Research.* New York, Free Press, 1976.
Lidz, Charles W., 'The Cop-Addict Game: A Model of Police Suspect-Interaction,' *J. Pol. Sci. Adm.* 2(1), 1974.
Mauss, Marcel, *The Gift.* Glencoe IL, Free Press, 1954.
Mayberry-Lewis, David, *The Savage and the Innocent.* Boston, Beacon, 1965.
Miller, S. M., 'The Participant Observer and 'Overrapport',' *American Soc. Rev.* 18, 97–99, 1953.
Parsons, Talcott, 'Evaluation and Objectivity in the Social Sciences: An Interpretation of Max Weber's Contributions,' *Deutsche Gesellschaft für Soziologie and the UNESCO Journal of the Social Sciences* 17(1), 1975.
Post, Emily, *Etiquette.* New York, Funk and Wagnells, 1945.
Roadburg, Allan, 'Breaking Relationships with Research Subjects: Some Problems and Suggestions,' in W. Shaffir, R. R. Stebbins, and Alan Turowetz, *Field Work Experience: Approaches to Social Research.* New York, St. Martn's Press, 1980.
Seidel, John and Jack Clark, 'The Ethnograph,' *Qualitative Sociology* 7 (1&2), 110–125, 1984.
Simmel, Georg, *The Sociology of Georg Simmel,* Kurt Wolff, editor. Glencoe, IL, Free Press, 1950.
Schwartz, Morris and Charlotte G. Schwartz, 'Problems in Participant Observation,' *Amer. J. Soc.* 60(4), 343–353, 1955.
Wax, Rosalie, 'Field Methods Techniques: Reciprocity as a Field Technique,' *Human Organization* 11(3), 34–37, 1952.
Weber, Max, *On the Methodology of the Social Sciences.* Glencoe, IL, Free Press, 1949.
Zerubavel, Eviatar, *Patterns of Time.* Chicago, University of Chicago Press, 1979.

DAVID SILVERMAN

TELLING CONVINCING STORIES:
A PLEA FOR CAUTIOUS POSITIVISM IN CASE-STUDIES

One way of describing the qualitative-quantitative distinction in the social sciences is by a fairly conventional sociology of knowledge. Viewed from the eastern edge of the Atlantic Ocean, one may detect two very different fashions in social research. Viewed by the contents of its learned journals, North America looks like the home of quantitative research. Survey research and, to a lesser extent, official statistics comprise the staple diet of published papers. The occasional case-study does appear but it seems to be self-conscious in such company and will often make every effort to dress itself up in the anonymous style of the scientific paper in order to make itself look less out of place.

Conversely, since the late 1960s, quantitative sociology has been on the defensive in British sociology. The successive waves of ethnomethodology (Garfinkel: 1967) and post-structuralism (Silverman and Torode, 1980) have had a significant impact, particularly on younger teaching staff and graduate students. Even without these theoretical challenges, field research had flourished on a fertile soil where, ironically, the under-funding of social science research had encouraged small-scale, observational work.

Now, of course, one must not exaggerate these odd British tendencies. First, as I later note, published work in the British journals still has its fair share of quantitative research, reflecting the continued strength of the much-abused 'positivism' among the older generation of scholars. Second, 'qualitative' research can cover a vast range of research styles and can even be co-opted back into the positivist tradition. For instance, in British market research circles, I gather that 'qualitative' research is the latest fashion. It is seen to provide 'in-depth' material which is believed to be absent from survey research data. Above all, it is relatively cheap. However, all that is meant by qualitative research is open-ended interviews or 'panel' studies. Finally, if British sociology is perceived to have a strength by North American social scientists, it lies in *neither* qualitative *nor* quantitative research but in social theory (think of the reputations of, say, Anthony Giddens or Charles Taylor).

Nevertheless, there remain real differences of emphasis on each side of the Atlantic. Qualitative social scientists represent, I suspect, a small, somewhat

Barry Glassner and Jonathan D. Moreno (eds.),
The Qualitative-Quantitative Distinction in the Social Sciences. 57–77.
© 1989 *by Kluwer Academic Publishers.*

beleaguered minority in North America, while the powerful critiques of the quantitative tradition which they produce (Cicourel, 1964, Schwartz and Jacobs, 1979), have, ironically, had more impact in Britain than at home. In Britain, it is the quantitative researchers who are anxiously looking over their shoulders, while attempting to cope with the new fashions that are influencing their graduate students (e.g. Brown, 1973).

Now the problem with fashions is that they often stand in opposition to critical thinking: 'party lines' fit more easily with sloganising than with original work. The message I derive from this observation is that the polarities around which the qualitative/quantitative distinction have been based need (to use the fashionable term) to be deconstructed. Why should we assume, for instance, that we have to choose between qualitative and quantitative methods? Why can we focus on only 'meanings' but not 'structure' or on 'micro' but not 'macro' processes? Why should case-study researchers assume that there is something intrinsicially purer in 'naturally-occurring' data?

This was the point I developed in a recent text on methodology (Silverman, 1985). The new generation of British social scientists, I feel, need to be rather less smug about the rectitude of their affirmed belief in a non-positivistic research programme. Programmes are no substitute for lateral thinking *and* rigour. It is also possible that North American qualitative researchers need to be rather less defensive about the logic of case-study work. In both cases, the problem is that the prevailing fashion establishes a rhetoric which stands in the way of critical thought.

Three final points before I move on to the substance of this chapter. First, this is a curious example of a programmatic statement of an anti-programmatic position: As I have already noted, the perceived strength of British sociology is its ability to make *ex cathedra* statements about the social world. Of course, this strength is also a weakness. What one might call the disease of *didactica nervosa* manifests itself in the desire to make inward-looking programmatic statements rather than to venture out from the academy to report on the outside world. Like the addict, I can only hope that one more fix will be the last that will be needed!

Second, I have tried to make the programmatic flavour of this chapter a little less dry by relating my arguments to a body of substantive literature. Much of this is drawn from the area of the sociology of health and illness. This is unashamedly selected as the field in which I have the most competence. I hope that the majority of my readers, who are unlikely to have this specialised interest, will bear with me. The research studies chosen are only important because they illustrate the research strategies I wish to discuss.

Finally some words of explanation may be in order about my title. The problem is that the two bits of the title seem to work in opposite directions. What has story-telling to do with cautious positivism? After all, sober researchers are expected to deal in evidence and data. They are not supposed to spin yarns.

The problem is compounded by the problematic character and size of the audience who might by sympathetic to my 'plea'. After all, isn't *most* of the work that gets done in the sociology of health and illness quite happy to be identified as 'cautious positivism'? Think, for instance, of the epidemiological work and of all those surveys into doctors' attitudes and patient satisfaction.

There is a real danger, then, of preaching to the converted, or of trying to convince a small, beleaguered minority (of non-positivistic researchers) that they ought to sell out. It all depends, however, on how that overworked term 'positivism' is being used.

ON 'POSITIVISM'

I will try to clarify the nature of (and the audience for) my 'positivist' programme. When I call for cautious positivism, I am *not* calling for a number of practices that have come to be associated with the term. In particular, I *reject*:

1. Basing all research on experimental data, official statistics or the random sampling of populations.
2. Only treating *quantified* data as valid or generalisable.
3. Assuming that the basic issue with research instruments is of a technical nature. So in interviews, for instance, we must ensure a standardised format to allow for comparability.
4. Having a cumulative view of data drawn from different contexts so that, as in trigonometry, we are able to *triangulate* the true state of affairs by examining where the different data intersect.
5. Having a stimulus-response model of change in which objective 'experts' engage in social engineering by manipulating variables to achieve desired correlations.

Each of these practices has a number of defects, many of which are discussed (and some displayed) in a number of texts concerned with qualitative methodology ranging from Cicourel (1964) through Denzin (1970) to Schwartz and Jacobs (1979) and Hammersley and Atkinson (1983). Following the same order as in the list above, I note that:

1. Experiments, official statistics and survey data may simply be inappropriate to some of the tasks of social science. They exclude the observation of 'naturally-occurring data' by means of ethnographies and other forms of case-study.

2. While quantification may *sometimes* be useful, it can conceal as well as reveal social processes. Consider the problems of counting 'attitudes' in surveys. Do we all have coherent attitudes on any topic awaiting the researcher's question? And how do 'attitudes' relate to what we actually do – our practices? Or think of official statistics on cause of death compared to studies of the socially organised 'death-work' of nurses and orderlies (Sudnow, 1967) and of pathologists and mortuary attendants (Prior, 1985).

3. Technical questions about research instruments conceal analytic issues. Do even 'standardised' interviews tell us simply 'facts about the world'? If so, what kind of facts? As will become clear shortly 'telling convincing stories' is, I believe, one of the most intriguing characteristics of interview-responses (see also Silverman, 1985, Chapter 8).

4. Triangulation of data seeks to overcome the context-boundedness of our materials at the cost of analysing their sense in context. For purposes of social research, it may not be useful to conceive of an overarching reality of which data, gathered in different contexts, are simply an index. Elsewhere, I have pointed to the inconsistency of those interactionists who seem to overlook their stated focus on 'situated actions' when there is an opportunity to triangulate data (Silverman, 1985, Chapter 5).

5. I believe that social engineering is élitist. Rather than seek to create brave new worlds, we might try to be more modest. I don't believe we downplay the role of social research if we seek instead to facilitate the circumstances under which people can make their own choices – create their own new worlds.

So this is no call for an uncritical acceptance of the standard recipes of conventional methodology texts or the standard practices of much research, whether or not they use the term 'positivism' (I suspect not because 'positivism' now has its home firmly in a grammar of abuse). Moreover, qualitative, case-study researchers have no reason to be so apologetic about their work – as if they might have done better to have looked at a larger sample. People doing case-studies should not assume that a quantitative logic must apply to their research. There is no reason for such researchers to be downcast when told that their work is not representative.

This has been well argued in an imporatant paper by Clyde Mitchell (1983) from which I have adapted the following table:

TABLE I

The logic of case-studies

	Survey research	Case studies
Claim to validity	Depends on representativeness of sample	Only valid if based on an articulated theory
Nature of explanations	Statistical and logical inference correlations not causes	Logical/causal connections
Relation to theory	Theory neutral	Theory dependent

As Mitchell points out, in survey research the claim to validity arises through using modern sampling methods to ensure that there is no bias in the selection of the sample. Similarly, in experimental studies, double-blind random control trials may be essential. However, in qualitative research, we choose a case because we believe it exhibits or tests some identified general theoretical principle. For instance, I was interested in fee-for-service medical consultations because of a more general concern with ceremonial orders (Strong, 1979). A study of medical practice in a 'private' clinic could thus broaden generalisations about the ceremonial order in NHS medicine, as well as highlight any cultural universals in medical practice as a whole (see Silverman, 1984).

The nature of explanation relates to the way we claim validity. Mitchell stresses that, in survey research, the issue is always the inference we make from the sample to the parent population. Our explanations solely relate to the concomitant variation of two or more characteristics. The *interpretation* of these correlations must depend on ungrounded theoretical musings about causation.

Conversely, the process of inference from case-studies is *only* causal and
cannot be statistical. The argument derives from the unassailability of the
analysis, not from the representativeness of the case.

This is why the validity of case-study analysis depends upon the adequacy
of the underlying theory. What we are doing in case-studies, Mitchell argues,

TABLE II.A

Cultural definitions of anomaly

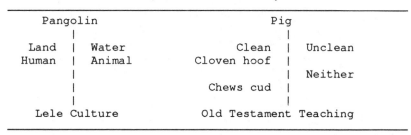

TABLE II.B

Cultural responses to anomaly

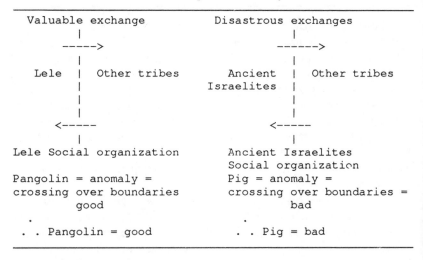

(Based on Douglas, 1975)

is seeking to specify the *necessary* conditions among a set of *theoretically-significant* elements. Unlike survey-research, we are not satisfied with explanations of most of the variance in our data. We seek to explain *all* our data, pursuing deviant cases as opportunities to refine our analysis so that the relationships we describe have been analytically induced from the data.

A brilliant example of this is provided in Mary Douglas' (1975) case-study of a Central African tribe, the Lele. Douglas was struck by the importance of an ant-eater called a pangolin in tribal ritual. The Lele perceived the pangolin as highly anomalous because, in their terms, it had some features of both land and water animals as well as possessing certain human characteristics (for instance it gave birth to only one offspring at a time).

Now, other tribes also remark on anomaly. For instance, in the Old Testament, the pig is singled out because it is both clean and unclean (see Table II.A). Yet, for the Ancient Israelites, this rendered the pig untouchable. Conversely, the Lele celebrated anomaly – in the form of the pangolin.

The cultural response to (perceived) anomaly provided the theoretical rationale for Douglas' case-study. Using the comparative method, she sought to specify the necessary conditions among her set of theoretically-defined elements. The Lele's 'deviant' response to anomaly could be grounded upon their experience of valuable exchange – across boundaries – with other tribes. The pangolin, which crossed the symbolic boundaries of cultural classification, was thus a symbolic representation of social relations. Conversely, the Israelites frowned upon the pig since it represented their bad experience of moving across boundaries (see Table II.B).

ON ANECDOTALISM

Unfortunately, not all case-study work possesses the elegance of Mary Douglas' research on the Lele. It is one thing to deploy damaging critiques of positivism. Altogether more complex intellectual resources are required to mount a convincing alternative.

Indeed, I detect an increasing dissatisfaction with the anecdotal character of some case-studies. There is a growing feeling that we should seek to go beyond producing nuggets of interesting information and seek to build generalisations. Non-positivist sociologists must really do something about their odd relation to facts – their temptation to refer always to 'facts'. As Strong (1985) has pointed out, there are a lot of facts around without inverted commas around them. The real question is deciding what are the *important* facts and working out what we are going to *make* of them.

Yet we have to accept that the fashion in the seventies, following the justified critiques of positivism, pushed us a long way from facts. Researchers forgot Weber's recognition of the problematic character of meaning and his reference to the state of 'inarticulate half-consciousness' in which most action takes place. Indeed, they got involved in a morass of 'subjectivity' and 'authenticity'. Few people were struck by the parallels between this kind of work and the dominant forms of representation on the media, with their emphasis on the personalisation of events.

Today times are harder for the population as a whole as well as for researchers. It is hardly surprising in this cruel new world that sociologists should think again about their research imperatives. For instance, Elliot Freidson has criticised subjectivism in the sociology of medicine:

Positivist assumptions have done considerable less damage to the value of sociological research than have the assumption of other groups. Some have replaced the myth of objectivity with a myth of subjectivity, holding that subjective reality *is* sufficient for reliability and validity and ignoring, even arguing against, formal methods (Freidson, 1983, 214).

In the different context of the sociology of education, Martyn Hammersley, comes to the same conclusion:

We have a wealth of theoretical ideas in sociology but a paucity of theories. The theoretical ideas we employ are rarely developed into an explicit and coherent form, and even more rarely are they subjected to systematic testing. This neglect of the development and testing of theory stems in large part from what, in my view, has been an over-reaction to 'positivism' (Hammersley, 1985, 244).

As Jennifer Platt (1981) has noted, when discussing 'positivism' it is far easier to make statements of principle 'rather than observing the complexities and ambiguities of practice' (85). Despite the theoretical critiques of positivism in the 1970s, Platt found little evidence of change in the kinds of papers published in recent sociology journals. Quantitative data and surveys continued to be used to the same extent, while there was a small increase in reports of research which sought to test hypotheses.

Platt concluded that there is no clearly defined 'positivist' style of research. Perhaps, she suggests:

all convincing empirical work is necessarily positivistic, whatever the researcher may claim about the matter in the abstract (Platt, 1981, 81).

Platt's attempt to find a common logic in 'all convincing empirical work' complements the increasing awareness of the weaknesses of some contemporary alternatives to positivism.

In a recent book, Hammersley and Atkinson (1983) have properly criticised the fashion of limiting research to the study of 'meanings' in naturally occurring settings. The *naturalism* has, they say, at least three problems:

1. The notion of 'natural' and 'artificial' settings is misleading. Artificial settings, set up by researchers, are still part of society and so display fundamental social processes. To which I would add that no data is purer or less contaminated than any other. Methods of analysis can be good or bad; data is ultimately just data (albeit constructed via a research strategy).
2. Naturalism seeks 'to tell it like it is', avoiding theory. Yet this is self-evidently mistaken. We only see the world through theories and through methods for acting upon it. Indeed, as I have just noted, the strength of case-study work should be its theoretical elegance.
3. Naturalism limits itself to cultural description. It only wants to describe how members see things. Yet, as Dingwall (1981) has noted, this task is literally endless, as well as *ad hoc*. Moreover, as the work on the Lele shows, really interesting questions only begin to arise when we introduce a comparative perspective.

Just as qualitative methodology should reach beyond the purely descriptive, there is some evidence that quantitative work is moving beyond the merely statistical. As Catherine Marsh (1979) has cogently argued, the best quantitative work does not maintain that causation can be established merely through a technology of correlations. Increasingly, such work attends to the contexts in which data arise and does not assume an 'individual true value', any deviation from which is due merely to 'response error'.

This suggests to me that we should retain what was worthwhile in older, perhaps more positivistic research traditions, aiming to develop cumulative bodies of knowledge built on falsifiable generalisations. Necessarily, this will mean that we will need to attend to facts as well as to 'facts'. It also implies that we cannot forget questions of policy and practice – the impact of facts upon people.

TELLING CONVINCING STORIES

This neatly leads me on to the issues referred to in my sub-title. These have clearly-defined analytic and methodological angles. I will discuss each in turn.

1. *'Convincing' as an analytic*: In the previous discussion I did not mean to downgrade the analysis of the social construction of 'facts'. Indeed, expressed in shorthand, one of the facts that constrain people is that many facts are 'facts'. This commonplace sociological observation has recently been given an interesting new twist in Prior's (1985) discussion of the categorisation of death. Cause of death is defined in a legal frame of reference – what Prior calls 'policing the dead' – but is then the basis of statistics which will be used in a medical context (epidemiology).

So, telling a convincing story, is to produce a tale that works because it relies on a narrative which is reflexively related to the conditions of its production. As Gubrium *et al.* have put it:

... description is constrained by membership obligations, not the least of which is that description is also a job – a material condition. To interpret good description as objective representation is to assume that one can, in principle, ignore the conditions of its production, both cognitive and material (Gubrium *et al.*, 1982, 19).

Gubrium is relying on a version of the character of description which has its home not only in relatively recent versions of ethnomethodology (Garfinkel, 1967) but also in the mainstream sociology of C. Wright Mills. His neglected paper on 'vocabularies of motive' shows that descriptions of human conduct appeal to motives which do not *derive* from the individuals as described but are imputed to them (Mills, 1940). He goes on to suggest a programme for considering when motive-talk gets done and the socially-organised resources used in motive attribution – for instance the basis of imputations of religious or hedonistic motives.

Similarly, Gubrium *et al.* call for a study of:

The concrete conditions of descriptive activity ... the diverse objects, resources, events, and social relations that both penetrate and serve as the working background of representations (Gubrium *et al.*, 1982, 24).

As they note, such social relations specify who is obligated to describe whom and the form and consequences of these descriptions. For instance, they point out that medical decision-making centres around two considerations:

the real problems and care of patiens as the staff concretely experience them, and how the experiences are to be presented to whomever staff is accountable (Gubrium *et al.,* 1982, 29).

As in Prior's work, Gubrium is concerned with accounts tied to their conditions of production. Gubrium shows elsewhere how such apparently solid social institutions as the family can be analysed as rhetoric – 'a way of socially ordering everyday life, and of course, influencing it at the same time' (Gubrium and Lynott, 1985, 132). Consequently, where people are perceived to show concern for one another, staff may refer to 'fictive' family as a way of describing social relations of patients at long-stay health institutions (Gubrium and Buckholdt, 1982).

Now this may seem to be a rather eccentric and highly abstract way of discussing 'convincing stories'. However, I believe it has an important methodological implication for those of us troubled by naturalism's demand to 'tell it like it is'.

No longer do we need to treat interview data as providing a privileged entry into the sacred realm of subjective experience. As Cain and Finch (1981) put it:

subjects' accounts are objects of study. They are not self-validating and they are not sociological explanations (110).

The aim, then, becomes that of analysing the account as a form of discourse or rhetoric. The rhetoric is convincing precisely because it reflects members' artful methods in appealing to culturally-based knowledge of reality.

In this way, we can dissolve much of the distinction between research-derived data and 'naturally occurring' data. A concern with rhetoric makes social organization equally accessible to the interviewer and the ethnographer.

Say we are concerned with family relations and either cannot obtain access or are troubled by the 'halo effect' if we do. If we follow Gubrium's argument, the problem vanishes:

An investigator need not attempt to get inside or intimate with families in order to get 'real life' understandings of what they actually are like. That aim is based on the assumption that family actualities are located and found somewhere within the confines of the concrete families proper. But where are these concrete social forms? ... Family rhetoric shows that the family at home is no more actual than the family anywhere. The study of family order turns on its representation, on its signs, and those whose rhetoric attempts to reveal it for what the latter take it surely to be (Gubrium and Lynott, 1985, 150). See also Silverman (forthcoming).

2. *'Convincing' as a methodology*: The value in methodological terms of treating rhetoric in this way is indisputable. However, unless we are to face an infinite regress, we must also address how to tell convincing stories about convincing stories. In doing research, we can convince others (and ourselves) by seeking to satisfy three criteria:

(a) Adopting research strategies which contribute worthwhile, theoretically based generalisations drawing from case-study work.

(b) Avoiding the kind of polarised thinking we have seen in the positivism/materialism debate. This includes those other either-or choices which so bedevil research – for instance the concern with meaning versus structure or qualitative versus quantitative methods (see Silverman, 1985, Chapters 2 and 7).

(c) This means accepting certain methods of science while rejecting many of its functions in society, particularly as social engineering or surveillance. (I will return to this point later).

Now, to some extent, these strictures may apply less to the sociology of health and illness than to other sociological sub-speculations. Our area has been less buffeted by fashionable theories than some others. Conversely, the movement away from a reliance on survey research and official statistics has been rather later to arrive here.

A little survey of the recent literature brought this home to me. Working backwards from May 1985, I looked at the eleven most recent issues of *Sociology of Health and Illness* and at the articles labelled 'medical sociology' in the sixteen most recent issues of *Social Science and Medicine*. Table III gives a breakdown of the data-base used in these papers.

These figures have to be viewed with caution. In particular, the sections entitled literature reviews/debates or essays include some material that I might have labelled case-studies. For instance, SHI contained a number of historical papers that I chose to label 'essays'. This was based on a quite arbitrary decision to limit the 'case-study' label to ethnographies and the like.

According to these tabulations, SSM is one-half survey research. If we include the 19% of macro description (ranging from accounts of Indian primary care to a description of the training of ambulance staff in the U.K.) the 12% based on official statistics and the 2% using content analysis, fully 82% of the papers use conventional 'positivist' methods.

Conversely, as might be predicted, only 21% of SHI is devoted to such methods (all survey-research studies), while the proportion based on case-study work (even using my limited definition) increases from 8% to 35%. Again, content analysis is strangely neglected in SHI but 7% of the papers use conversation analysis. Overall, 42% of SHI papers use 'non-positivist' methods as against 8% in SSM.

TABLE III

Contents of two journals

Social Science and Medicine (1984–5) (SSM)

	Number of papers	Percentage of whole
Survey research	54	49%
Macro description	21	19%
Official statistics	13	12%
Literature reviews/debates	11	10%
Case studies	9	8%
Context analysis	2	2%
	110	100%

Sociology of Health and Illness (1982–5) (SHI)

	Number of papers	Percentage of whole
Survey research	13	21%
Literature reviews/essays	22	37%
Case studies	21	35%
Conversational analysis	4	7%
	60	100%

I will now limit myself to the case-study papers and discuss the criteria through which we might assess whether they told convincing stories. I should emphasize that the avoidance of 'positivism' is by no means the key to success. Indeed, the 'hands-off' attitude towards official statistics revealed by the contributors to SHI means that a potentially valuable source of data went neglected – although medical records were discussed in interesting ways by Prior's (1985) paper on death certificates and in Heath's (1984) work on G. P.'s notes.

For this assessment, I shall use three criteria which I touched upon earlier:

1. Building cumulative bodies of knowledge
2. Based upon falsifiable generalisations
3. Being concerned with questions of policy and practice

1. *Cumulative bodies of knowledge:* Here I must add that not *any* body of knowledge will do. Obviously, we can accumulate all the knowledge we like about, say, the relation between left-handedness and behaviour without developing any *sociologically* worthwhile generalisations.

A simple 2×2 table may illustrate the kind of research I have in mind here.

TABLE IV

Research, history and type of knowledge

| | | Scope of study | |
		Generalisation	Reportage
Basis	Theory-based	Exchange and anomaly	Instances of 'labelling' or 'practical reasoning'
	Theory-neutral	Voting behaviour	'Anthropological tourism'

The table gives *ad hoc* examples of the kind of work which fits each of the cells. As will be obvious, I prefer research which fits into the top left-hand corner of the table. Again, using Douglas' (1975) study of the Lele, the best kind of work can use case-studies to develop theory-based generalisations. Her paper does this by:

(a) Developing *formal* theories from elsewhere (Lévi-Strauss, 1955, discussion of the role of anomaly as a mediating entity in systems of classification).

(b) Linking micro and macro relations, by moving from the cultural particulars of the Lele to global symbolic organisation.

(c) Using lateral thinking which crosses substantive and conceptual boundaries. For instance, Douglas seeks to understand the Lele's own lateral thinking by relating it to what is depicted in cave-dwellers' paintings. The latter also seem to celebrate anomalous entities.

Some of these strengths are found in the papers which I have been reviewing. I have set out some examples in Table V, although I should stress that the selection of papers is purely for illustrative purposes. I do not intend to put a seal of approval on certain work, nor to cast other papers into outer darkness.

TABLE V

Examples of theory-based generalisations in case-studies

Developing formal theories

Epileptics interviewed: patient control rather than compliance (Conrad SSM)
Long-term treatment institutions: conditions when rules applied consistently (Roth SSM)
Doctor-patient relations: trials and coalitions (Pendergast SHI)
Admissions to mental hospitals: re-defining 'normality' (Mestrovic SHI)
Hospital encounters: 'sentimental work' (Strauss SHI)
Medical records: creating patient careers (Heath SHI)
Accident departments: rules of clinical priority (Dingwall and Murray SHI)

Linking macro and micro relations

Mortality statistics: practices of mortuary workers and coders (Prior SHI)
Pap smears: medical dominance and hidden agendas (Fisher SHI)

Lateral thinking

Comparing occult and conventional programmes for disease (Davies SSM)
Accounts of causes of arthritis as 'narrative reconstructions' (Williams SHI)

In some other cases, however, little attempt was made to develop theory-based generalisations. Theory seemed just to play a legitimating role: it was stated at the outset and then forgotten. It would be invidious to identify examples of such a common failing. It is sufficient to cite a study of time-perceptions which made no attempt to feed into theories of social time (e.g. Bourdieu, 1977) and an examination of forms of adaptation in 'total institutions' which merely *cited* Goffman. Other papers mentioned Glaser and Strauss' (1967) discussion of 'grounded' and 'formal' theories but then did not develop any formal theories.

Overall, 11 out of 21 case-study papers in SHI attempted some form of theory-based generalisation – a bare majority. The same was true of only 3 out of 10 papers in SSM.

2. *Attempts at Falsification*: Speculation can use as much lateral thinking as it likes but it will achieve little unless it is well-grounded in data and attempts are made to falsify generalisations.

Over two decades ago, Howard Becker and Blanche Geer (1960) discussed how research might produce unexpected findings. They suggested:

(a)   Trying to distinguish the necessary from the sufficient conditions of some phenomenon.
(b)   Careful inspection of all deviant cases.
(c)   Distinguishing statements made to the observer alone from those made to others in everyday conversation.
(d)   Counting instances where appropriate.

Becker and Geer's second suggestion parallels the attention to deviant cases in the methods of analytic induction recommended by Mitchell (1983). This method is a way of confronting the two basic issues that face all research (see Smith: 1975):

Validity: does the account offer a watertight, logically-based interpretation of the data? Can it withstand plausible, rival interpretations?
Reliability: will the same approach and methods, used by different researchers, produce the same results?

Adequacy at the level of theory is insufficient unless we can trust the reliability of the interpretation. A recent paper by Marshall (1985) has set out some additional criteria for reliability or what she calls 'trustworthiness'. These include:

(a)   Preserving and displaying the data
(b)   Showing and explaining negative instances
(c)   Making data collection and analysis 'semi-public not magical' (355)
(d)   Keeping fieldwork logs
(e)   Distinguishing between 'site-specific findings and findings generalisable to other settings' (356).

Now it is true, as Marshall points out, that a single-minded search for trustworthiness could undermine the value of qualitative research, losing the richness of its data in exchange for an empty rigour. She shows that the 'tests in contexts' recommended by Huberman and Miles (1983) for reliable qualitative research may simply reduce it to a pale carbon copy of survey research. Nonetheless, one can be more confident of the validity and reliability of qualitative work where attention is paid both to Mitchell's call for rigorous theorising and to Becker, Geer and Marshall's criteria for trustworthiness.

It is sad to report that very few applications of such criteria were found in the case-studies I examined. One such example was Dingwall and Murray (1983). They demonstrate that being a 'bad' patient – assumed to be responsible for your own illness and unco-operative – is insufficient to explain medical staff's rules of clinical priority. It is possible to be neither a 'good' nor a 'bad' patient but simply 'legitimate but routine.' In order to distinguish the necessary conditions of clinical priority, they focus on a deviant case – children. Children are of theoretical interest because, although they are likely to have the features of 'bad' patients, they will not be defined as such by doctors.

Sue Fisher's (1984) paper on the micro-politics of a gynaecological clinic displays a similar logic. In order to discuss the social circumstances in which a pap-smear is obtained, she distinguishes a decision-making 'tree' and advances her analysis through deviant cases. Finally, Lindsay Prior (1985) has neatly shown the relation between the way in which the causes of death are ascertained and counted and the social organisation of the medico-legal system. There is no trace of irony in his analysis: the statistics are relevant not because they show 'bias' but because they are the necessary outcome of socially organised practices.

However, such attention to the kind of criteria discussed by Becker and Geer was largely confined to these three papers. The much more typical pattern was displayed in assertions based on no more than supportive gobbets of data. For instance, according to a study of student doctors:

> 'Two such cases seemed illustrative'
> 'As one intern noted...'
> 'Intern Charen's comments were typical' (Mizrahi, 1985)

In SSM, despite the scientific pretensions suggested by its name, no case-study I looked at showed any visible attempt at rigour. This no doubt reflects the fact that SSM concentrates on quantitative work, while being prepared to accept a few case-studies on the strength of their insights rather than their reliability. The unfortunate consequence is that far from coming up with unexpected data, we tend to find papers confirming what researchers, journalists or indeed anybody might have guessed already. For instance, one paper related drug-use and punk life-style on unashamedly anecdotal evidence (Burr, 1984).

3. *Questions of policy and practice*: I suppose academic journals might not be the place one would write if one were attempting to influence real life situations. Therefore one should not be surprised to find here an absence of discussions of conventionally-defined action research. Nor is any attention paid in these papers to the possible contribution of research to local communities or local struggles or to the possibility of dialogue with the people studied.

So, although overall, a proportion of 4 out of 5 papers reviewed did touch upon the policy implications of their research, all of these limited themselves to a version of knowledge as enlightenment. Somehow, in unknown ways, the studies were expected to sensitize the people concerned to the issues raised. To summarize:

(a)   Social engineering was certainly avoided – but at the cost of little involvement in the real world. Based on a predictable political reformist outlook, the overall sense that these papers conveyed was 'isn't it terrible?' or 'there they go again'.

(b)   Authors tended to adopt a facile Fabian version of all knowledge as good. The way in which professionalised forms of knowledge and surveillance already control our lives was conveniently overlooked.

(c)   Apart from one or two papers sympathetic to feminism (e.g. Fisher, 1984), little attention was given to the relation between research and actual struggles in the real world.

If I had to make a plea for more involvement in these issues, I would suggest three modest proposals.

(a)   Attention to reforming structures to allow people themselves to innovate – the *opposite* of social engineering.

(b)   Involvement and dialogue with people concerned with struggles at all stages of the research process.

(c)   A rejection of the administrative mentality which, when wedded to reformism, wants to make more and more of human life subject to surveillance by psychological and social experts. As my own experience has taught me, the most well-meaning questioning of such diverse groups of adolescents as diabetics (Silverman, 1985a) and cleft-palate patients (Silverman, 1983) usually reverts into a psychological interrogation. Why cannot health professionals limit themselves in such cases to a purely enabling role, offering help and advice as requested? The extension of the medical gaze implied in reformist, whole-person medicine may, unintentionally, increase surveillance around and between subjectified bodies (see Armstrong, 1983 and Silverman, forthcoming).

CONCLUDING OBSERVATIONS

It has not been my intention to award marks to the work of my peers. This outbreak of *didactica nervosa* does not arise from any position of unques- tioned authority. Much of what I have had to say could be turned back, in a critical fashion, on my own work, partly because any insights I have are derived *ex post facto* from my own intellectual biography.

My central theme has been that, if qualitative research has failed to live up to its promise, it is because polarised thinking has prevented us from a more critical approach to data. Rigour is not out of place in qualitative research and the claims of the comparative method and of deviant-case analysis are as strong as ever. Nor need we worry about using interview material or simple qualitative data in case-studies, providing that we have thought through the sometimes naive assumptions of 'triangulation'.

In all such case, I suspect that fear of anomaly has made qualitative research far less exciting than it might be. Perhaps we could turn Mary Douglas' (1975) analysis back on ourselves. If, unlike the Lele, we exist in an unfavourable environment, where our experience of exchange with other groups has been unrewarding, is it any surprise that we should fear anomaly and construct strong boundaries between schools of thought.

In a more practical sense, as we negotiate with our paymasters and clients, we might remind ourselves of two things:

1. We have no need to be ashamed of case-study work. There is nothing sacred about random-sampling or numbers.
2. This does not give us licence to be merely anecdotal. So we must seek cumulative generalisations that are theoretically derived *and* refutable. And a good place to start might be to think about the analytics and methodology of 'telling convincing stories'.

REFERENCES

Armstrong, D. (1983), *The Political Anatomy of the Body*, Cambridge: Cambridge University Press.

Becker, H. S. and Geer, B. (1960), 'Participant Observation: The Analysis of Qualitative Field Data', in Adams, R. M. and Preiss, J. J. (eds.), *Human Organization Research: Field Relations and Techniques*, Homewood, Ill.: Dorsey Press.

Bourdieu, P. (1977), *Outline of a Theory of Practice*, Cambridge: Cambridge University Press.

Brown, G. (1973), 'Some Thoughts on Grounded Theory', *Sociology 7* (1), 1–16.

Burr, A. (1984), 'The Ideologies of Despair: A Symbolic Interpretation of Punks' and Skinheads' Usage of Barbiturates', *Social Sciences and Medicine 19* (9), 929–38.

Cain, M. and Finch, J. (1981), 'Towards a Rehabilitation of Data' in Abrams, P. *et al., Practice and Progress: British Sociology 1950–80*, London: Allen & Unwin.

Cicourel, A. (1964), *Method and Measurement in Sociology*, New York: Free Press.

Denzin, N. (1970), *The Research Act in Sociology*, London: Butterfield.

Dingwall, R. (1981), 'The Ethnomethodological Movement', in Payne, G., Dingwall, R., Payne, J. and Carter, M. (eds.), *Sociology and Social Research*, London: Croom Helm.

Dingwall, R. and Murray, T. (1983), 'Categorisation in Accident Departments: 'Good' Patients, 'Bad' Patients and Children', *Sociology of Health and Illness 5* (2), 127–48.

Douglas, M. (1975), 'Self-Evidence', in *Implicit Meanings*, London: Routledge & Kegan Paul.

Filmer, P. *et al.* (1972), *New Directions in Sociological Theory*, London: Collier-Macmillan.

Fisher, S. (1984), 'Doctor-Patient Communication: A Social and Micro-political Performance', *Sociology of Health and Illness 6* (1), 1–29.

Freidson, E. (1983), 'Viewpoint: Sociology and Medicine: A Polemic', *Sociology of Health and Illness 5* (2), 208–219.

Garfinkel, H. (1967), *Studies in Ethnomethodology*, Englewood Cliffs, N.J.: Prentice-Hall.

Glaser, B. and Strauss, A. (1967), *The Discovery of Grounded Theory*, Chicago: Aldine.

Gubrium, J., Buckholdt, D. and Lynott, R. (1982), 'Considerations on a Theory of Descriptive Activity', *Mid-American Review of Sociology VII* (1), 17–35.

Gubrium, J. and Buckholdt, D. (1982), 'Fictive Family: Everyday Usage, Analytic, and Human Service Considerations', *American Anthropologist 84* (4), 878–85.

Gubrium, J. and Lynott, R. (1985), 'Family Rhetoric as Social Order', *Journal of Family Issues 6* (1), 129–52.

Hammersley, M. (1985), 'From Ethnography to Theory: A Programme and Paradigm in the Sociology of Education', *Sociology 19* (2), 244–59.

Hammersley, M. and Atkinson, P. (1983), *Ethnography: Principles in Practice*, London: Tavistock.

Heath, C. (1984), 'Participation in the Medical Consultation: the Co-Ordination of Verbal and Non-Verbal Behaviour between Doctor and Patient', *Sociology of*

Health and Illness 6 (3), 311–38.

Huberman, A. and Miles, M. (1983), 'Drawing Valid Meaning from Qualitative Data: Some Techniques of Data Reduction and Display', *Quality and Quantity 17* 281–339.

Lévi-Strauss, C. (1955), 'The Structural Study of Myth', *Journal of American Folklore 78*, 270, 428–44.

Marsh, C. (1979), 'Problems with Surveys: Method and Epistemology', *Sociology 13* (2), 293–305.

Marshall, C. (1985), 'Appropriate Criteria of the Trustworthiness and Goodness for Qualitative Research on Educational Organizations', *Quality and Quantity 19*, 353–73.

Mills, C. W. (1940), 'Situated Actions and Vocabularies of Motive', *American Sociological Review 5*, 904–13.

Mitchell, J. C. (1983), 'Case and Situational Analysis', *Sociological Review 31* (2), 187–211.

Mizrahi, T. (1985), 'Getting Rid of Patients', *Sociology of Health and Illness 7* (2), 214–235.

Platt, J. (1981), 'The Social Construction of 'Positivism' and Its Significance in British Sociology: 1950–80', in Abrams, P. *et al., Practice and Progress: British Sociology 1950–80*, London: Allen & Unwin, 73–87.

Prior, L. (1985), 'Making Sense of Mortality', *Sociology of Health and Illness 7* (2), 167–190.

Schwartz, H. and Jacobs, J. (1979), *Qualitative Sociology: A Method to the Madness*, New York: Free Press.

Silverman, D. (1983), 'The Clinical Subject: Adolescents in a Cleft-Palate Clinic', *Sociology of Health and Illness 5* (3), 253–74.

Silverman, D. (1984), 'Going Private: Ceremonial Forms in a Private Oncology Clinic', *Sociology of Health and Illness 18* (2), 191–202.

Silverman, D. (1985), *Qualitative Methodology and Sociology*, Aldershot: Gower.

Silverman, D. (1987), *Communication and Medical Practice: Social Relations in the Clinic*, London and Beverly Hills: Sage.

Silverman, D. (forthcoming), 'Six Rules of Qualitative Research: A Post-Romantic Argument', *Symbolic Interaction*.

Silverman, D. and Torode, B. (1980), *The Material Word: Some Theories of Language and Its Limits*, London: Routledge.

Smith, H. W. (1975), *Strategies of Social Research: The Methodological Imagination*, London: Prentice-Hall.

Strong, P. (1979), *The Ceremonial Order of the Clinic*, London: Routledge & Kegan Paul.

Strong, P. (1985), Seminar paper, Department of Sociology, Goldsmiths' College.

Sudnow, D. (1967), *Passing On: The Social Organization of Dying*, Englewood Cliffs, N.J.: Prentice-Hall.

Williams, G. (1984), 'The Genesis of Chronic Illness: Narrative Reconstruction', *Sociology of Health and Illness 6* (2), 175–200.

DAVID J. SYLVAN

THE QUALITATIVE-QUANTITATIVE DISTINCTION IN
POLITICAL SCIENCE

In the School of political Projectors I was but ill entertained; the Professors appearing
in my Judgment wholly out of their Senses; which was a Scene that never fails to
make me melancholy.[1]

So great is the force of laws, and of particular forms of government, and so little
dependence have they on the humours and tempers of men, that consequences almost
as general and certain may sometimes be deduced from them, as any which the
mathematical sciences afford us.[2]

It goes without saying that one can only arbitrarily assign dates to intellectual
movements. Let us therefore be circumspect, and content ourselves with the
following observations. Swift's satire of the grand Academy of Lagado was
first published in 1726; Hume's essay, in 1741. Apparently, by the early 18th
century, 'political science' had become a recognizable object of discourse in
England. One might treat it with seriousness or scorn, but there was in any
case an 'it' that could be discussed.

To a remarkable degree, 'political science' has kept many of the contours it
displayed 250 years ago. Specifcally, it is a 'science' practiced (at least
partly) by 'Professors' who aim at 'deducing' 'general' 'consequences' as
'certain' as those of 'mathematics.' This certainty may be facilitated, if not
guaranteed, by means of various 'Methods; (Swift's term), of which some are
more useful than others. Indeed, if one abstracts from the term 'method', it
seems as if Swift and Hume long ago adumbrated the essential characteristic
not only of contemporary political science, but of other contemporary social
sciences.

This statement is halfway true. Modern political science does indeed
conform to the desiderate put forward by Hume (and satirized by Swift); so
too do subfields in other social sciences. However, there are major segments
of the latter which escape 18th century dicta in a way which political science
does not. That asymmetry is the subject of this paper.

Specifically, I intend to do three things. First, I will argue that political
science is one-sided. It focuses on questions of magnitude while ignoring
those dealing with meanings, even though both issues are studied in other
social sciences.[3] Second, I will trace this disjunction between political

79

Barry Glassner and Jonathan D. Moreno (eds.),
The Qualitative-Quantitative Distinction in the Social Sciences. 79–97.
© 1989 *by Kluwer Academic Publishers.*

science and other social sciences to the emergence of modern political science and sociology at the University of Chicago in the 1920s. Finally, I will examine the figure of Harold Lasswell to see just how political science failed to grapple with meanings.

1. POLITICAL SCIENCE TODAY

To get an idea of how the discipline of political science shapes up today, it is instructive to look at a year's worth of articles in the leading journal of the profession, the *American Political Science Review* (APSR). Somewhat arbitrarily, I have chosen the four issues that make up Volume 80 (calendar year 1986). The results are instructive.

1.1. Quantitative and Qualitative Methodologies

For the sake of discussion, I will use the term 'quantitative methodologies' to refer to research designs in which magnitudes play an essential part. Thus, 'quantitative methodologies' include the use of arguments resting on the results of procedures from fields in which key concepts are deemed to be capable of being arrayed along a scale of greater to less. Examples of such fields are:

Probability theory;
Descriptive statistics;
Differential calculus; and
Game theory.

It may seem perverse to lump together such disparate fields of knowledge, or to suggest that there is any mileage to be gained in using the same term to characterize both formal mathematical models and empirical data analysis. What does it matter, after all, if research designs assume that concepts can meaningfully be analyzed in terms of their magnitude?

I cannot pretend to answer this question fully here, but I can suggest at least one response to it. For a concept to be considered as validly analyzed by means of magnitude-based techniques, it must be considered as being essentially quantifiable. In other words, it is a constitutive feature of the concept that it can be discussed in terms of magnitudes (whether ordinal, integer, or ratio); could one not discuss the concept in this way, the concept would in effect be different. Any such quantifiable concept must perforce stand in a relation of independence to the pragmatic use (meaning, in the

Wittgensteinian sense) of the concept by the participants to whom it pertains. They may in fact use it magnitudinally, but they need not. In either case, the essential quantifiability of the concept is guaranteed by the analyst, and justified on grounds (e.g., power, elegance) separate from the participants' use. In anthropological language, quantitative methodologies are etic rather than emic; in Cicourel's (1964) terminology, they are the *locus classicus* of measurement by fiat.

I hasten to add that etic research designs are not restricted to those based on quantitative methodologies. Many types of legal analysis, for example, are etic, as are standard journalistic methodologies. But these latter are not intrinsically etic (q.v. discussions of intentionality in criminal law), while quantitative methodologies are. The irrelevance of participants' meanings is presupposed in any quantitative design, for the simple reason that magnitude assignments are by definition monotonic, and therefore syntactically invariant. For example, a concept such as 'intensity of preference' may vary greatly by relation to some other concept (e.g., 'type of candidate', 'time'), but by definition higher intensities are always greater than medium intensities, while the latter are always greater than low intensities. This monotonicity is of necessity syntactically invariant: no matter the semantic content that attaches to 'high' or 'low', it is always the case that higher intensities are greater than low, and that it is a solecism to speak of lower intensities being greater than high ones.

For participants, however, concepts cannot be so circumscribed. Any notion is used in a potentially inexhaustible range of contexts, and the use varies with context. For example, a politician may be said by his peers to be strong, but the way in which they use the word, and the types of comparisons they draw between strength and weakness, vary dizzyingly by spatio-temporal context. Thus, we can imagine it being said of A that his strength weakens him by comparison with B, who is weak, when it comes to vote-trading. Strength is not always stronger than weakness; it depends on the context. Now, admittedly, we do use the word 'weakens' in a syntactically invariant fashion, but there are so many contexts in which it can be used that it can be applied to almost any referent. In short, from a participant's point of view, a concept cannot be syntactically invariant; and it follows that magnitude assignments by nonparticipants are intrinsically etic.

These considerations shed some light on the term 'qualitative methodologies': it refers to research designs in which participants' categories ('qualities') are preserved. This is not to say that such designs are perforce *limited* to participants' categories, but simply that additional concepts must preserve the meanings participants attach to their terms. For example, scholars employing qualitative methodologies may work toward the elaboration of conceptual schemas encompassing the range of uses that participants make of their own terms (cf. Spradley, 1979),[4] indeed, if meanings were shared widely in a particular subculture, one could even imagine the use of descriptive (though not, I would guess, inferential) statistics to map out that culture's meanings. Such mapping-out, however, is quite distinct from the 'accounting-for' at the heart of the qualitative enterprise.

This last point bears expansion. In order to preserve meanings, it may in fact be useful to represent certain uses in formal, as well as illustrative, way. To return to the political example discussed above, we might find it useful to represent the range of meanings by a grammar that specifies (and, with examples, depicts) the 'proper' use of particular terms in particular contexts (e.g., Polanyi, 1985). Other formal modes of representation that preserve participants' meanings may also be imagined. On the argument above, none of those modes would be quantitative. (It is important not to confuse formalism with quantification. There are any number of formal methodologies [e.g., predicate logic; transformational grammar; point-set topology] which do not presuppose magnitudes; conversely, it is possible to employ quantitative reasoning without being formal [e.g., statements linking 'more' of something to 'less' of something else].) But they would clearly be 'rigorous' (i.e., explicit and systematic), just as can also be the case for the exploratory methodologies (e.g. participant observation, ethnographic interviewing) more commonly associated with qualitative work.

As should by now be apparent, I am using the terms 'quantitative methodology' and 'qualitative methodology' in an oppositional manner. The opposition in question is, of course, the emic-etic distinction, which is itself recognized (albeit at times dimly) by researchers in both (qualitative and quantitative) traditions. But in fact, the opposition between qualitative and quantitative methodologies goes a good deal deeper. In spite of their shared empiricist core (Sylvan and Glassner, 1985), adherents of each methodology define their activities in opposition to those of the other. To see how this works, we may now return to the APSR.

1.2. Literature Exploration

What is most striking about examination of volume 80 of the APSR – at least in light of the discussion above – is the almost total absence of qualitative work. Of 41 full-fledged articles, only one (Dietz, 1986) employs any kind of qualitative methodology. Significantly, it is a hermeneutic analysis of what Machiavelli really meant to do in *The Prince*; Machiavelli's meaning in the book is interrogated by reference to his other writings, his political activities, and an esthetic of Renaissance pictorial art. Every other article but one in the volume is either a quantitative study, or an advocacy of such; or else it is a kind of journalistic account of what particular authors (specifically, Freud, Thucydides, Foucault, Socrates, Hegel, Aristotle, and Marx) wrote about some subject and why we might be interested in it today.

The one exception is a piece by Fenno (1986) advocating a type of participant observation for studying the United States Congress. The article – in fact Fenno's APSA presidential address – consists of a series of anecdotes about Senators told to illustrate the details one finds when engaged in 'observation'; in that respect, it echoes Fenno's earlier panegyrics on participant observation when applied to Representative (1978). However, even though Fenno's work is replete with hundreds of detailed observations, its aim is essentially journalistic: to provide enough details of a story that complex theoretical regularities can be postulated and verified.[5] Such detailed piecing-together – who said (or wrote) what to (or about) whom, when, and why – is the core of journalistic activity. It may be carried out with common journalistic techniques – interviews, summaries or exegeses of documents – or, as in Fenno's case, by quasi-sociological ones, but its etic core remains the same.[6] Of the articles in volume 80 of the APSR, eight are either exercises in journalism or advocacy of it. Except for the Dietz article, the rest are quantitative.

This ratio – one qualitative study to thirty-two quantitative ones – is not uncommon in political science. A search through other major journals of the discipline reveals similar ratios. Book listings among major publishers show somewhat lower ratios (though still very few qualitative studies), with correspondingly greater emphasis on journalistic studies. Research methods textbooks in the discipline either ignore qualitative methodologies entirely (e.g., Kweit and Kweit, 1981; Boynton, 1980), or include a token chapter unconnected with the rest of the book (e.g., Manheim and Rich, 1981, ch. 11). Graduate courses in 'scope and methods' or research design show similar disproportions. Even scholars highly critical of standard quantitative methodologies (e.g., Sartori *et al.*, 1975) typically fail to articulate a qualitative alternative. In short, it is no exaggeration to conclude that qualitative methodologies are close to absent from political science.

They are, of course, not completely absent. Certainly, Collingwood-type studies of historical meaning exist (e.g., Pitkin, 1972); the Dietz article

(1986) in the APSR falls into this category. Then, too, there is the occasional work in the Geertz mode of 'thick description' (e.g., Scott, 1976, 1985). But the majority of studies concerned with qualitative methodology are no so much instances of it as they are calls for it. Thus we find a small but hardy subliterature criticizing quantification and advocating analyses of meaning (e.g., Bernstein, 1978; MacIntyre, 1971; Taylor, 1971) alongside of a methodological and programmatic literature that sketches out what qualitative studies would look like (e.g., Alker, 1975, 1976, 1979, 1982; Edelman, 1977, 1985). Even the form of these propaedeutic works – defensive or hortatory – is evidence of just how sparse the qualitative terrain is in political science. In other social sciences, one finds methodological primers about how qualitative work is commonly carried out (e.g., Schwartz and Jacobs, 1979); such a primer would be impossible to write in political science for the simple reason that there is relatively little qualitative work to write about.

In spite of the enormous disproportion between quantitative and qualitative work in political science, there is a sense in which the two approaches are rival siblings. As we have seen, the majority of works concerned with qualitative methodology are principally concerned with setting forth just why quantification is bad and what ought to be done instead. Were it not for the ubiquity of quantitative work, most writings in the qualitative perspective would have no *raison d'être*. By the same token, qualitative methodology appears to quantitative scholars as a messy or solipsistic limit that defines the boundaries of what can rigorously be understood. For example, a classic work of political science, Robert Dahl's *A Preface to Democratic Theory*, is permeated by distrust about relying on individuals' reports of their feelings or concerns; the latter are considered 'sensations' that 'social science has not yet caught up with' (1956, 102). Similarly, another of the discipline's classic works, David Mayhew's *Congress: The Electoral Connection*, postulates reelection as every Representative's proximate goal, adding in a tone of relief 'It will not be necessary here to reach the question of whether it is possible to detect the goals of congressmen by asking them what they are, or indeed the question of whether there are unconscious motives lurking behind conscious ones' (1974, 16n).

We thus arrive at a curious conclusion. On the one hand, qualitative methodologies in political science are close to an empty set. On the other hand, they act as the limits of the discipline, defining both semiotically (through certain buzz words like 'rigor' or 'subjective') and praxiologically just what it is that most political scientists want not to do. But to state the matter in this way raises another question. We may take it as a truism that any discipline has boundaries which 'discipline' its members. Why, though, are the boundaries in political science those of qualitative methodology rather than some other discipline, such as biology, journalism, or law? One could certainly make the case (and it has, at times, been made) that there is some core to political science which is defined in opposition to how practitioners in these or other fields do or could study politics. There is nothing intrinsic to qualitative methodologies which makes them, in effect, the bugaboos of contemporary political science. To understand how this came to be, we shall have to take a backward glance at the emergence of political science in its modern disciplinary form.

2. THE CHICAGO SCHOOL

By almost any criterion, political science assumes its modern form at the University of Chicago in the 1920s. Charles Merriam became chairman of the political science department there in 1923, and immediately embarked on a process of recruiting and institution-building. Within a decade, Merriam's efforts led to the discipline's being recast in a form which persists to our day. Consider three aspects of political science as we know it today.[7]

2.1. Merriam's Legacy

First, we can look at institutional characteristics. In the 1920s, Merriam helped to establish the Local Community Research Committee at the University of Chicago. Through his connections with private foundations, the Committee soon became well-funded, so much so that it could afford to underwrite both faculty and graduate student research. With adequate funding support, scholars were able to use their new training in statistics and what came to be known as survey research, and put it to work in large-scale, empirically ambitious studies. A model of joint, collaborative research began to be established. At the same time, the hands-on experience of conducting surveys or gathering other forms of data led to the development and refinement of new quantitative techniques.

Merriam also was instrumental in the establishment of other research facilities. He was a major figure in setting up the Social Science Research Council, which came on a national level to be what the Local Community Research Committee was in Chicago. Both groupings were avowedly interdisciplinary, providing an institutional venue for the importing into political science of methodologies and concepts from other social sciences.

Merriam even helped obtain Rockefeller Foundation money to build a Social Science Research Building at Chicago. It is fair to say that Merriam worked out the institutional solutions to how major quantitative studies could be done in political science: the use (and hence training) of graduate students; the use of university and national fund-granting committees; and the regularized reliance on outside foundations for support. Except for the existence of government funding, contemporary institutional mechanisms for quantitative research in political science adhere to the structures established by Merriam some 60 years ago.

The second aspect of the discipline that Merriam helped establish in modern form was the reproduction of the Chicago School in a number of other major universities. The transmission mechanism, of course, was the students trained at Chicago by Merriam and his colleagues, who then went on to academic careers in other institutions. I cannot even begin to give a complete listing of all the lines of affiliation,[8] but will simply note several names: Gabriel Almond, who played a pivotal role in shaping the department at Stanford; Harold Guetzkow, who did the same at Northwestern; V. O. Key, who cast the Harvard department in its modern form; and David Truman, who helped transform the Columbia department. In turn, the students of these scholars spread the word to other places. Of course, there were many other Chicago students (and, as we will see below, faculty members) whose ideas played a vital role in the establishment of the discipline (e.g., Harold Lasswell, Herman Pritchett, and Herbert Simon), but their contributions to the formation of specific 'prestige' departments were more diffuse. Suffice it to say that the Chicago model of research and graduate student training, as well as the research agendas set out over half a century ago, are alive and well in most major political science departments today.

But my argument hinges on the establishment of a specific intellectual orientation – quantitative methodology – and it is on this third aspect of the discipline that I now wish to concentrate. Throughout the 1920s, Merriam called persistently for the use of quantitative methodologies in political science research. In 1921, 1922, and 1923, he spoke before the APSA on the subject, served on research committees of the Association established to explore that question, and published five articles on the topic in the APSR and the *National Municipal Review*. His arguments were collected and expanded in *New Aspects of Politics* (1925), which is essentially a manifesto for quantitative methodologies. Its chapter titles are worth noting: 'The Foundations of the New Politics' (about science); 'Recent History of Political Thinking' (about the growth of quantitative work); 'Politics and Psychology';

'Politics and Numbers'; 'Politics in Relation to Inheritance and Environment'; 'Political Prudence' (about data collection); 'The Next Step in the
Organization of Municipal Research' (about large-scale surveys); and 'The
Tendency of Politics' (about new research projects that could be undertaken).[9] Throughout *New Aspects*, the tone is one of quiet confidence, as one
recent study after another is adduced as examples of what Merriam has in
mind. This is a far cry from the pessimistic assessment of only five years
before that in the first two decades of the twentieth century 'the influence of
modern scientific method [was] relatively weak' (1920, 431).

Merriam set about recasting political science into a quantitative enterprise.
With his student (later colleague) Harold Gosnell, he wrote *Non-voting*
(1924), a work employing random sampling, descriptive statistics, and a
large-scale survey. Students provided the bulk of the interviewers; punch
cards and counter-sorters were used to tabulate the results. Merriam and
Gosnell followed up this collaboration by increasingly quantitative revisions
of Merriam's textbook on American political parties (Merriam, 1922;
Merriam and Gosnell, 1929, 1940, 1949).

Although Merriam himself was not a methodologist, he in fact wrote the
methodological chapters of *Non-voting*. Nonetheless, Merriam's chief
contributions to the development of quantitative methodologies in political
science lay in the hiring he did at Chicago as chairman of the department.
One scholar he hired was Leonard White, whose work on public administration (1929) provided the survey research on that topic. Another Merriam
appointee was Quincy Wright, whose massive study of war (1965; begun in
the 1930s) employed a wide variety of quantitative methodologies. But a
special place must be reserved for those former graduate students at Chicago
whom Merriam later hired for the faculty of the department. One was
Frederick Schuman, whose influential textbook on international politics (in
successive revisions, it stayed in print for some 30 years) is replete with both
quantitative data and magnitudinal claims (1933). Another student-turned-
Chicago-professor was Merriam's collaborator, Gosnell, whose heavily
statistical studies look as if they could have been published yesterday (1927,
1937). The most famous such in-house appointee, of course, was Harold
Lasswell, who will be dealt with later in this essay. The point, however,
should be clear: whatever the quality of Merriam's own work, he was a
superb judge of intellectual quality, and both his appointees and his students
(often the same person) transformed the research programmes of political
science into something quite close to their present state.[10]

I do not mean to suggest that modern political science was solely a creation of Merriam and his colleagues. Quantification was clearly 'in the air' in the 1920s, a fact recognized by both advocates and opponents of Chicago-style research (e.g., respectively, Rice, 1928; Laski, 1926). Those at Chicago clearly did not invent quantitative political science out of whole cloth; some magnitude-based methodologies for the analysis of politics had been around for over 50 years. Nor were the Chicago scholars the only ones to engage in quantitative research. But, as I have argued above, they took those methodologies and incorporated them into their everyday research and graduate training activities, thereby constituting the discipline as a distinctive combination of workplace styles and intellectual agendas. In a very real sense, they invented the discipline as we know it today. If we wish to answer the question of how political science came to be overwhelmingly, hegemonically quantitative, we have only to look back at Merriam and his associates.

2.2. The Other Chicago School

In spite of the above account, a mystery persists. For political science was not the only discipline to get defined in its modern form at Chicago in the 1920s. The same can be said of sociology, but in even stronger terms: 'Chicago School' has a resonance in the latter field that it never developed in the former. We know that members of both departments collaborated on both university and research activities, and that graduate students in one department took courses in the other. Yet the intellectual orientation of Chicago sociology was vastly different than that of Chicago political science.

Specifically, the Chicago School of sociology developed both qualitative and quantitative methodologies, whereas its political science counterpart concentrated solely on the latter. A detailed genealogy of Chicago sociology is outside the ambit of this paper; but some of its way stations may be usefully recalled.[11] Albion Small was the first chairman of the department; through his translations of Georg Simmel's writings, students at Chicago were early on exposed to a particular type of qualitative methodology (Kantian formalism). One of Small's colleagues at Chicago (in the field of social psychology) was George Herbert Mead, whose theories of the act and of the self (e.g., 1934) were enormously influential in laying out a conception of meaning that lies at the heart of one of the major qualitative paradigms, symbolic interactionism. Mead's lectures were avidly attended by a large number of students, many of whom later went on to become prominent scholars in their own right. But his ideas apparently were also influential on his faculty contemporaries, such as his colleague in sociology, W. I. Thomas.

By 1920, the groundwork had been laid for much qualitative work in sociology.

In the 1920s and 1930s, sociologists at Chicago began to develop the fieldwork techniques that became a hallmark of qualitative methodology. From the outset, those techniques were understood as contributing to analyses of participants' meanings. Thus, in his introductory essay to a collection of urban fieldwork monographs, Robert Park stated that 'The city is, rather, a state of mind, a body of customs and traditions, and of the organized attitudes and sentiments that inhere in these customs and are transmitted with this tradition' (1925, 1). Park's colleague and fellow urban sociologist, Louis Wirth put the issue more programmatically:

Whereas in dealing with the objects in the physical world the scientist may very well confine himself to the external uniformities and regularities that are there presented without seeking to penetrate into the inner meaning of the phenomenon, in the social world the search is primarily for an understanding of these inner meanings and connections ... Hence insight may be regarded as the core of social knowledge. It is arrived at by being on the inside of the phenomenon to be observed, or, as Charles H. Cooley put it, by sympathetic introspection. It is the participation in an activity that generates interest, purpose, point of view, value, meaning, and intelligibility, as well as bias (Wirth, 1936; cf. Wirth, 1928).

By the late 1920s, scholars at Chicago had begun to broaden the range of their qualitative studies to include the study of language. Robert Redfield, an anthropologist affiliated at times with the sociology department, used fieldwork techniques to investigate the routine grammar of particular terms. Significantly, Redfield's preface to *Tepoztlan* (1930) thanks two teachers: the linguistic anthropologist Edward Sapir, and the sociologist Robert Park. Herbert Blumer, a student and later professor at Chicago, brought some of these linguistic concerns back into sociology, as yet another way of using Mead's ideas in sociological inquiry (Blumer, 1969). Hence, in a variety of ways, Chicago sociology in the 1920s and 1930s was responsible for the constitution of a major strand of modern qualitative sociology.[12]

While this work went on, other sociologists at Chicago were engaged in the development of quantitative methodologies. Chief among them was William Ogburn, who worked extensively on the development of social indicators. Interestingly, Ogburn wrote the foreword to Gosnell's *Machine Politics*; he also was responsible for the motto (from Lord Kelvin) chiseled on the Social Science Research Building: 'When you cannot measure, your knowledge is meagre and unsatisfactory.'[13] Other of Ogburn's colleagues (e.g., Ernest Burgess), played important roles in the development of quantitative sociology.

The Chicago School of sociology thus appears as far more pluralistic than its counterpart in political science. If practitioners of qualitative methodologies did not always see eye to eye with their quantitative colleagues, they nevertheless coexisted, and helped constitute the discipline as encompassing both types of research. Given the close connections between faculty and students in both Chicago departments, the obvious question arises: why sociology and not political science?

Any answer to this question must of necessity be speculative. It is notoriously difficult to prove a negative, and even to begin to answer the question, one would have to delve into details of the intellectual biographies of some dozen individuals. Lacking both the space and the resources to pursue that track, I shall not pursue it further. Rather, I propose to turn to the one Chicago political scientist who could, more than any of his colleagues, have served as a founder of qualitative political science. I refer, of course, to Harold Lasswell. If we cannot answer the question of *why* political science failed to develop a thriving qualitative wing, we can at least discover *how* it was that Lasswell failed to do so.

3. THE TRAGEDY OF HAROLD LASSWELL

By all accounts, Lasswell was a polymath. His range of learning, his acquaintance with enormous numbers of disparate literatures, was extraordinary. To a number of his interlocutors, he appeared to have read everything. Rosten, for instance, gives an awed account of a Lasswell ad lib performance in which, among other authors, he discussed in detail Marx, Jung, Freud, Adler, Radek, Wolfe, Klein, Lukacs, Thurstone, Sherrington, Henderson, Carlson, Lao Tse, Sapir, Tawney, Weber, Heisenberg, Schrodinger, Rousseau, Catlin, Merriam, Malaparte, Sorel, Dicey, Sombart, Keynes, Bryce, Whitehead, Lenin, Eddington, Lewis, Boas, Michels, and Spengler (Rosten, 1969, 8).[14] This command of literature was coupled with a fantastically inventive mind, one which led to pioneering work in the analysis of elites, of development, of communications (along with inventing much of content analysis), of personality dynamics, and of policy-making.

Early in his career, Lasswell recognized the importance of participants' meanings. As a student, he studied with Mead and, through him, became deeply interested in Whitehead's philosophy. Lasswell's doctoral dissertation posed, but did not attempt to answer, the question of the precise means by which propaganda could be 'transmitted into the experience-world of the subjects' (1927, 211). Within a few years, he had worked out a tentative methodology to explore those means: 'prolonged interviews with individuals under unusually intimate conditions' (1951 [1930], 204). In that study, *Psychopathology and Politics*, Lasswell used such interviews to analyze several types of politically active individuals.

For our purposes, though, what is most noteworthy about *Psychopathology* is its final chapter, a Meadian *tour de force* in which Lasswell argues that the state should be considered as 'a time-space manifold of similar subjective events' (245).[15] Part-way through the chapter, Lasswell addresses the problem of voting, pointing out that its meaning varies considerably across persons, places, and times. In a footnote, he writes that 'Perhaps the most important part of the Merriam and Gosnell study of *Non-voting* is the symposium of conversational scraps which suggest what ballotting actually means to various classes and sections of a modern metropolitan community' (249n). This point is followed by an advocacy of field-ethnographical methodology to investigate politics, along with a claim that psychopathological techniques of depth interviewing serve to complement the findings of the field ethnographer. In making these latter statements, Lasswell sounds very much like a modern anthropologist or sociologist. 'The whole aim of the scientific student of society is to make the obvious unescapable' (250). Quantitative methods are worthwhile only as a starting point for qualitative analyses of life histories (251–252). The quantitative approach is 'impressionistic' and involves 'simplification' of the complexities of meaning (252–254).

Thus by 1930, Lasswell had laid out a research agenda for qualitative research in political science. It called for a combination of interview and participant observation methodologies aimed at exploring how, for a variety of different types of individuals, political institutions such as the state were constituted from the meaningful activities in which those individuals were engaged. However, instead of following up on this research programme, Lasswell undertook to recast it into a large-scale, international survey, in which individual or group meanings would be subordinated to aggregate data on national personality types. Although the manifesto for this project (1935) was as scintillating intellectually as *Psychopathology*, it was a profound change from the latter. Instead of studying what symbols meant from an emic viewpoint, Lasswell would study them from an etic viewpoint. The shift can be seen clearly in his well-known *Politics: Who Gets What, When, How* (1936). In that book, symbols are presumed by Lasswell to have certain meanings; no serious effort is made to see whether those meanings are shared by the members of a particular cultural grouping.

Once Lasswell had made the shift, he never looked back. By the time of his large-scale RADIR studies in content analysis (1949), he completely reversed the terms of his discussion in *Psychopathology* about qualitative and quantitative methods. Rather than see the latter as fit for preliminary work, because of their simplifying etic characteristics, Lasswell came to see them as rigorous and systematic. Rather than see the former as desirable because of their ability to explore participants' meanings, Lasswell came to see them as *ad hoc* and sloppy. Not surprisingly, these are the very criteria held up by modern quantitative methodology in political science. Later pronouncements by Lasswell on the qualitative-quantitative distinction in political science (e.g., 1961) repeated those criteria, albeit with a somewhat more jazzed-up lexicon.

How did Lasswell come to shift his research agenda? A clue may be found in his 'general framework' essay (1948 [1939]). Lasswell begins the piece with some of his standard Meadian remarks about meaningful acts, then ties them to personality systems. The latter, in turn, are tied to cultural context, and it is argued that if one wishes to understand the first elements in that chain, one must needs study the widest possible contexts in which they are situated. Hence, the desirability of large-scale data surveys (and, 20 years later, to the establishment of the Yale World Data Project). Now, the critical move in that argument, at least for purposes of this paper, is the connection of meaning to personality. If one's personality is the context within which acts are imbued with meaning, then the enterprise is coherent. But by viewing things in this way, Lasswell assumes that one's personality is in some sense independent of one's acts. (An alternative view, one more in line with Lasswell's earlier qualitative bent, would be to see personality as a manifold of acts, a constitutive product of meaningful activities.) Such independence is possible, for example, if personality is a composite of genetic traits, or memories, or beliefs: if, in short, personality is something internal against which the (external) act can be gauged.

By making this assumption, Lasswell turns Mead on his head. The constitutive relation between self and gesture that Mead so subtly teases out is flattened to a perceptual filtering relationship that Mead was at pains to argue against. The meaningful becomes the internal; psychopathological analysis becomes political psychology; interviews and observation become attitudinal surveys and galvanic skin responses. Similar flattening processes may be seen in Lasswell's other synoptic works (e.g., Lasswell and Kaplan

1950), with similar quantitative methodologies being advocated. 'Intensive research' – one of Lasswell's catchphrases – is to this day understood in the flattened form that Lasswell turned to in the later 1930s (e.g., Brown, 1974).

Beyond this point, we cannot go. I have no way of knowing how Lasswell came to make his intellectual shift; though I suspect it has something to do with the intellectual appeal of psychoanalysis, and with Lasswell's interest in addressing the 'world revolution of our time'. Qualitative work, after all, is spatio-temporally limited. But this is speculation, and cannot be argued for with any certainty.

What we can say is that Lasswell represented the best hope of constituting political science as both a quantitative and a qualitative discipline. Further, we can say that he glimpsed that possibility, but turned away from it. Perhaps Lasswell could not have succeeded in opening up a qualitative space in political science even if he had tried. His failure to try for more than a few years, though, contributed to the one-sidedness of the discipline. For those who treasure openness, it is indeed a tragedy.

NOTES

* Useful suggestions were made by James Farr, Robert Holt, and W. Phillips Shively.
[1] Jonathan Swift, *Gulliver's Travels*, pt. 2, ch. 6.
[2] David Hume, 'That Politicks may be reduc'd to a Science,' in *Essays Moral, Political, and Literary*.
[3] On the assumption that economics is a social science, it, too, fails to deal with meanings. That one-sidedness is outside the ambit of this paper; interesting arguments on related subjects may be found in Schumpeter (1954).
[4] Scholars need not agree with the participants' *definitions* of the terms in order to preserve participants' *meanings*. Thus, Glassner and Loughlin (1987) write about three different types of adolescent 'light' drug users, even though such types are not identical to those discussed by their informants. However, the abducing of each type is connected to the specific, variegated, and not always consistent terminologies employed by the informants. In this sense, participants' meanings are preserved.
[5] In fairness to Fenno, he uses language that seems to evince a qualitative bent: q.v., politicians as 'goal-seeking and situation-interpreting individuals' whom we can gain knowledge of 'by trying to see their world as they see it' (1986, 4). But in practice, Fenno does no such thing. In spite of their copious quotations, his participant-observation studies repeatedly discard participants' terminology: how they charac-terize what they are doing, thinking, and so on. Instead, we are presented with a series of bland theoretical terms (e.g., 'sequence'; 'timing') that are based on Fenno's (not the participants') terminology. Thus, the work is essentially etic and, in spite of Fenno's protestations, journalistic.
[6] On the etic nature of most journalism, see Sylvan and Glassner (1985).
[7] For the discussion which follows, I have drawn on several sources for accounts of

Merriam's activities. The most useful are a biography of Merriam (Karl, 1947); a reminiscence by Lasswell (notably, in a festschrift for Quincy Wright, 1971); a contemporary evaluation by Louis Wirth and other of Merriam's Chicago colleagues (1940); and an autobiographical essay by Merriam himself, which is in fact the first chapter of his festschrift (1942). By contrast, the standard history of the discipline (Somit and Tanenhaus, 1967) contains almost nothing of value except incomplete bibliographic lists.

[8] Between 1920 and 1940, 80 doctorates in political science were awarded at Chicago (Karl, 1974, 147).

[9] It is important to remember that for Merriam, there was no essential separation between political science research and active involvement in reform politics: both were, in effect, forms of political systematic and scientific political theory.

[10] The process fed back on itself in a second way. Merriam learned from his students, incorporated their ideas into his own books (e.g., 1934), and posed new research problems for the next generation of Chicago students and faculty.

[11] A useful compendium may be found in Martindale (1981).

[12] Not solely responsible, because a phenomenological strand entered more recently, via other institutions.

[13] The motto was not met with universal applause, even by quantitative scholars. Frank Knight, of the University of Chicago Economics Department, said at the tenth anniversary celebration of the Social Science Research Building, 'There has been a lot of foolishness connected with this attempt to measure everything. There is, for example, this statement of Lord Kelvin's carved outside the window of this building ... Its practical meaning tends to be: 'If you cannot measure, measure anyhow'' (Wirth, 1940, 169).

[14] For additional biographical material on Lasswell, see Eulau (1969); Marvick (1977); and Smith (1969).

[15] The terms 'manifold' and 'event' were chosen deliberately by Lasswell to point to his intellectual debt to Whitehead. For Lasswell's own testimony on this, see (1948, 195).

REFERENCES

Alker, Hayward (1975), 'Polimetrics: its descriptive foundations,' *Handbook of Political Science,* eds. F. Greenstein and N. Polsby, **8**: 139–210. Reading: Addison-Wesley.

Alker, Hayward (1976), 'Research paradigms and mathematical politics,' *Sozialwissenschaftliches Jahrbuch für Politik,* ed. R. Wildenmann, **5**: 13–50. Munich: Günter Olzog Verlag.

Alker, Hayward (1979), 'From information processing research to the sciences of human communication,' *Informatique et Sciences Humaines* **40–41**: 407–420.

Alker, Hayward (1982), 'Logic, dialectics, politics: some recent controversies.' *Poznan Studies in the Philosophy of the Sciences and Humanities,* ed. H. Alker, **7**. Amsterdam: Rodopi.

Bernstein, Richard (1978), *The Restructuring of Social and Political Theory.* Philadelphia: University of Pennsylvania Press.

Blumer, Herbert (1969), *Symbolic Interactionism: Perspective and Method*. Berkeley: University of California Press.

Boynton, G. R. (1980), *Mathematical Thinking about Politics: An Introduction to Discrete Time Systems*. New York: Longman.

Brown, Steven (1974), 'Intensive analysis in political research,' *Political Methodology* **1**: 1–25.

Cicourel, Aaron (1964), *Method and Measurement in Sociology*. New York: Free Press.

Dahl, Robert (1956), *A Preface to Democratic Theory*. Chicago: University of Chicago Press.

Dietz, Mary (1986), 'Trapping the prince: Machiavelli and the politics of deception,' *American Political Science Review* **80**: 777–799.

Edelman, Murray (1977), *Political Language*. New York: Academic Press.

Edelman, Murray (1985), *The Symbolic Uses of Politics*, 2nd edn. Urbana: University of Illinois Press.

Eulau, Heinz (1969), 'The maddening methods of Harold D. Lasswell: some philosophical underpinnings,' *Politics, Personality, and Social Science in the Twentieth Century: Essays in Honor of Harold D. Lasswell*, ed. A. Rogow. Chicago: University of Chicago Press.

Fenno, Richard (1978), 'Notes on method: participant observation,' Appendix to *Home Style: House Members in Their Districts*. Boston: Little, Brown.

Fenno, Richard (1986), 'Observation, context, and sequence in the study of politics,' *American Political Science Review* **80**: 3–15.

Glassner, B. and Loughlin, J. (1987), *Drugs in Adolescent Worlds: Burnouts to Straights*. London: Macmillan.

Gosnell, Harold (1927), *Getting Out the Vote: An Experiment in the Stimulation of Voting*. Chicago: University of Chicago Press.

Gosnell, Harold (1937), *Machine Politics: Chicago Model*. Chicago: University of Chicago Press.

Karl, Barry (1974), *Charles E. Merriam and the Study of Politics*. Chicago: University of Chicago Press.

Kweit, M. and Kweit, R. (1981), *Concepts and Methods for Political Analysis*. Englewood Cliffs: Prentice-Hall.

Laski, Harold (1926), *On the Study of Politics: An Inaugural Lecture Delivered at the London School of Economics and Political Science on 22 October 1926*. Oxford: Humphrey Milford/Oxford University Press.

Lasswell, Harold (1927), *Propaganda Techniques in the World War*. New York: A. A. Knopf.

Lasswell, Harold (1935), *World Politics and Personal Insecurity*. New York: Whittlesey House.

Lasswell, Harold (1936), *Politics: Who Gets What, When, How*. New York: McGraw-Hill.

Lasswell, Harold (1948), 'General framework: person, personality, group, culture,' *The Analysis of Political Behaviour: An Empirical Approach*. New York: Oxford University Press. Orig. publ. *Psychiatry* **2** (1939): 375–390.

Lasswell, Harold (1949), 'Why be quantitative?', *Language of Politics: Studies in Quantitative Semantics*, H. Lasswell, N. Leites et al. New York: George W.

Stewart.

Lasswell, Harold (1951), *Psychopathology and Politics*, in *The Political Writings of Harold D. Lasswell*. Glencoe: Free Press. Orig. publ. Chicago: University of Chicago Press, 1930.

Lasswell, Harold (1961), 'The qualitative and quantitative in political and legal analysis,' *Quantity and Quality*, ed. D. Lerner. Glencoe: Free Press.

Lasswell, Harold (1971), 'The cross-disciplinary manifold: the Chicago prototype,' *The Search for World Order: Studies by Students and Colleagues of Quincy Wright*, eds. A. Lepawsky, E. Buehrig, and H. Lasswell. New York: Appleton-Century-Crofts.

Lasswell, H. and Kaplan, A. (1950), *Power and Society: A Framework for Political Inquiry*. New Haven: Yale University Press.

MacIntyre, Alasdair (1971), 'Is a science of comparative politics possible?' *Against the Self-Images of the Age*. Notre Dame: University of Notre Dame Press.

Manheim, J. and Rich, R. (1981), *Empirical Political Analysis: Research Methods in Political Science*. Englewood Cliffs: Prentice-Hall.

Martindale, Don (1981), *The Nature and Types of Sociological Theory*, 2nd edn. Boston: Houghton Mifflin.

Marvick, Dwaine (1977), 'Introduction: context, problems, and methods,' *Harold D. Lasswell on Political Sociology*, ed. D. Marvick. Chicago: University of Chicago Press.

Mayhew, David (1974), *Congress: The Electoral Connection*. New Haven: Yale University Press.

Mead, George Herbert (1934), *Mind, Self, and Society from the Standpoint of a Social Behaviorist*, ed. C. Morris. Chicago: University of Chicago Press.

Merriam, Charles (1920), *American Political Ideas: Studies in the Development of American Political Thought, 1865–1917*. New York: Macmillan.

Merriam, Charles (1922), *The American Party System: An Introduction to the Study of Political Parties in the United States*. New York: Macmillan.

Merriam, Charles (1925), *New Aspects of Politics*. Chicago: University of Chicago Press.

Merriam, Charles (1934), *Political Power: Its Composition and Incidence*. New York: Whittlesey House.

Merriam, Charles (1942), 'The education of Charles E. Merriam,' *The Future of Government in the United States: Essays in Honor of Charles E. Merriam*, ed. L. White. Chicago: University of Chicago Press.

Merriam, C. and Gosnell, H. (1924), *Non-voting: Causes and Methods of Control*. Chicago: University of Chicago Press.

Merriam, C. and Gosnell, H. (1929), *The American Party System: An Introduction to the Study of Political Parties in the United States*, rev. edn. New York: Macmillan.

Merriam, C. and Gosnell, H. (1940), *The American Party System: An Introduction to the Study of Political Parties in the United States*, 3rd edn. New York: Macmillan.

Merriam, C. and Gosnell, H. (1949), *The American Party System: An Introduction to the Study of Political Parties in the United States*, 4th edn. New York: Macmillan.

Park, Robert (1925), 'The city: suggestions for the investigation of human behavior in the urban environment,' *The City*, by R. Park, E. Burgess, and R. McKenzie. Chicago: University of Chicago Press.

Pitkin, Hanna (1972), *Wittgenstein and Justice*. Berkeley: University of California Press.
Polanyi, Livia (1985), *Telling the American Story: A Structural and Cultural Analysis of Conversational Storytelling*. Norwood: Ablex.
Redfield, Robert (1930), *Tepoztlan, A Mexican Village: A Study of Folk Life*. Chicago: University of Chicago Press.
Rice, Stuart (1928), *Quantitative Methods in Politics*. New York: Alfred A. Knopf.
Rosten, Leo (1969), 'Harold Lasswell: a memoir,' *Politics, Personality, and Social Science in the Twentieth Century: Essays in Honor of Harold D. Lasswell*, ed. A. Rogow. Chicago: University of Chicago Press.
Sartori, G., Riggs, F., and Teune, H. (1975), 'Tower of Babel: on the definition and analysis of concepts in the social sciences,' *Occasional Paper* **6**. Pittsburgh: International Studies Association.
Scott, James (1976), *The Moral Economy of the Peasant: Rebellion and Subsistence in Southeast Asia*. New Haven: Yale University Press.
Scott, James (1985), *Weapons of the Weak: Everyday Forms of Peasant Resistance*. New Haven: Yale University Press.
Schuman, Frederick L. (1933), *International Politics: An Introduction to the Western State System*. New York: McGraw-Hill.
Schumpeter, Joseph (1954), *History of Economic Analysis*, ed. E. Schumpeter. New York: Oxford University Press.
Schwartz, H. and Jacobs, J. (1979), *Qualitative Sociology: A Method to the Madness*. New York: Free Press.
Smith, Bruce (1969), 'The mystifying intellectual history of Harold D. Lasswell,' *Politics, Personality, and Social Science in the Twentieth Century: Essays in Honor of Harold D. Lasswell*, ed. A. Rogow. Chicago: University of Chicago Press.
Somit, A. and Tanenhaus, J. (1967), *The Development of American Political Science: From Burgess to Behavioralism*. Boston: Ally and Bacon.
Spradley, James (1979), *The Ethnographic Interview*. New York: Holt, Rinehart and Winston.
Sylvan, D. and Glassner, B. (1985), *A Rationalist Methodology for the Social Sciences*. Oxford: Basil Blackwell.
Taylor, Charles (1971), 'Interpretation and the sciences of man,' *Review of Metaphysics* **25**: 3–51.
White, Leonard (1929), *The Prestige Value of Public Employment in Chicago*. Chicago: University of Chicago Press.
Wirth, Louis (1928), *The Ghetto*. Chicago: University of Chicago Press.
Wirth, Louis (1936), Preface to K. Mannheim, *Ideology and Utopia*, trans. L. Wirth and E. Shils. New York: Harcourt, Brace.
Wirth, Louis, ed. (1940), *Eleven Twenty-Six: A Decade of Social Science Research*. Chicago: University of Chicago Press.
Wright, Quincy (1965), *A Study of War*, 2nd edn. Chicago: University of Chicago Press. Orig. Publ. 1942.

STJEPAN G. MEŠTROVIĆ

SCHOPENHAUER'S WILL AND IDEA
IN DURKHEIM'S METHODOLOGY

There are many ways of apprehending the qualitative-quantitative distinction in social research, of course. A useful approach is to distinguish these two methodologies along the rough outlines of the object-subject debate, a distinction that has taken many forms in various arguments. Nevertheless, according to most textbooks, the quantifiers are supposed to be 'objective' and positivistic. They allegedly deal with 'facts' and their 'totem' is purported to be Emile Durkheim. By contrast, the qualifiers are supposed to pay attention to the actor's 'subjective' point of view, and they are said to be phenomenological.[1] Max Weber is often cited as their 'totem.' Much has been written about the epistemological crises in sociology that stem from pushing this object-subject distinction to an extreme.[2] Recently, Horowitz has exposed the ideological biases that afflict sociological theory as a result of this distinction.[3] Nevertheless, notwithstanding the many fine efforts that have been made to transcend this distinction, it is undeniable that it continues to inform – in some manner, however dilluted – contemporary social research textbooks, and that it continues to afflict sociology.

Plato was among the first in Western thought to insist that the most important aspect of an argument is not the logic that supports it but the starting point from which it begins. The starting point for the debates and discussions mentioned above seems to be that the founding fathers of sociology were positivistic. And positivistic methodologies definitely presuppose a radical distinction between object and subject. Yet, an important founding father of sociology, Emile Durkheim, argued *against* the positivistic splitting of object and subject – his starting point seems to have been something relatively new in philosophy. The Durkheimians advocated what Marcel Mauss called the concept of 'totality.'[4] Mauss' 'total social fact' assumes the simultaneous study of the social, psychological and physiological dimensions of phenomena.[5] According to Mauss, Durkheimian sociology 'presupposes the combined study of these three elements: body, mind and society.'[6] Mauss elaborates:

In reality ... rarely or even hardly ever ... do we find man divided into faculties. We are always dealing with his body and his mentality as wholes, given simultaneously and all at once. Fundamentally body, soul and society are all mixed together here ... [these are] the phenomena of totality.[7]

Barry Glassner and Jonathan D. Moreno (eds.),
The Qualitative-Quantitative Distinction in the Social Sciences. 99–118.
© 1989 *by Kluwer Academic Publishers.*

Mauss argues that the study of social phenomena as 'totality' was the goal of psychology and anthropology as well as sociology in his day – he even mentions Freud and Jung as two exemplars of this approach. He felt that the 'heroic age' in which his viewpoint had to be defended was over in his day, and that the social sciences could continue with this holistic task. Obviously, his optimistic conclusion was premature. Another one of Durkheim's followers, Célestin Bouglé, writes in a neglected essay that 'according to this conception sociology appeared less as a separate discipline occupied exclusively with the formal side of groupings ... than as a synthesis of particular social sciences.'[8] Bouglé insisted that 'the necessity of specialization must not make us lose sight of the ultimate ideal, which is to achieve a synthesis and to *renew* the various social sciences.'[9] What philosophy served as the starting point for this kind of social science, one that aimed at synthesis, 'totality' and the reconciliation of object and subject?

I believe the answer lies in Arthur Schopenhauer's philosophy. The influence of Schopenhauer's *The World as Will and Idea*[10] has been almost completely overlooked in the history of the social sciences. Typically, one comes across Comte, Saint-Simon and Kant in such analyses, but not Schopenhauer. Yet Schopenhauer was the rage in nineteenth century thought because he seemed to resolve Kant's dilemma of how the world could be known as more than mere phenomenon, and because he focused on what might be called the politics of desire. Auguste Comte's philosophy was, in contrast to Schopenhauer's, overly 'intellectualist,' as Bouglé[11] and Pierre Halbwachs[12] have put it. One will find Schopenhauer's two key terms, 'representation' and 'will,' verbatim or by implication, in the writings of many late nineteenth century writers. This is especially true for Durkheim and Freud. Durkheim's friend and colleague in the French Philosophical Society, André Lalande, remarked, upon the occasion of the celebration of the 100th anniversary of Durkheim's birth, that Durkheim was so enamored with Schopenhauer that his students nicknamed him 'Schopen.'[13] Durkheim's pivotal claim that society is a system of 'representations' constantly battling the 'bottomless abyss' of human desires is a faithful refraction of Schopenhauer's version of *homo duplex*.[14] Freud's many dualisms, all off which pit the 'higher' aspects against the 'lower' id-like aspects, are a more obvious, and more freely admitted, reflection of Schopenhauer. For example, Freud confessed that to a large extent 'psychoanalysis coincides with the philosophy of Schopenhauer.'[15]

The aim of this essay is to suggest a new starting point for commonplace debates and accepted ways of thinking about Durkheim's relevance to the distinction between qualitative and quantitative studies. By 'representations,' Schopenhauer was not referring to the representations *of* reality in the minds of human agents – a move that only exacerbates the object-subject distinction he sought to overcome – but to abstractions that behave like realities. By 'will,' Schopenhauer did not mean deliberate action, obligation, intentions or effort, but passion and desire as they work against or despite reason.[16] Qualitative methodologies that focus on the 'actor's subjective point of view' in contemporary sociology are still overly intellectualist relative to Schopenhauer's philosophy because they do not typically invoke the unconscious and other manifestations of the imperious will. And quantitative methodologies are still arrested in the Kantian dismissal of *noumena* as unknowable and unmeasurable. Based on the premises of Schopenhauer's philosophy, Durkheim seems to have attempted to combine aspects of what we now distinguish as quantitative and qualitative methodologies.

SCHOPENHAUER'S OPPOSITION BETWEEN THE REPRESENTATION AND WILL

From the perspective of making an argument, the choice of Schopenhauer as the context for this discussion is somewhat arbitrary – of course, all such choices are. But from the perspective of choosing a good starting point for subsequent arguments, it was made carefully. According to Magee, 'by the turn of the century ... Schopenhauer was an all-pervading cultural influence'[17] – consider Schopenhauer's influence on Nietzsche and Wittgenstein in philosophy, Freud in psychology, and a host of artists, among them Tolstoy, Conrad, Proust, Zola, Maupassant, Hardy, and Mann.[18] Thomas Mann[19] describes the rest of Schopenhauer's more unacknowledged, diffused but still powerful influence on turn of the century thought with distinguished eloquence – but he omits Durkheim and other sociologists. Yet Schopenhauer was apparently Durkheim's favorite philosopher, and Georg Simmel[20] wrote a book on Schopenhauer and Nietzsche as their thought pertains to the social sciences. Let us re-consider Schopenhauer's importance briefly, paying particular attention to sociology and the object-subject distinction.

  Both of Schopenhauer's pivotal concepts, representation and will, were pedestrian in the nineteenth century. Schopenhauer attempted to reform the insights of Plato, Kant, Hegel, and all the major philosophers prior to him with regard to the question 'What about the world is knowable?' It was Schopenhauer, not Comte, who dominated late nineteenth century philosophical thought. The French term 'représentaton,' which is translated into English as 'idea,' and its German equivalent 'Vorstellung,' were standard terms not only for the Durkheimians but also among the phenomenologists, Freudians and most philosophers.[21] According to Janik and Toulmin,[22] representations were discussed in the drawing rooms of Vienna and Paris for at least a century before Durkheim and Freud came on the scene. As a result of Kant's devastating critique of pure empiricism, thinkers at the turn of the century turned to the task of analyzing the basic data of experience as the structured phenomena 'representations.' Schopenhauer was the leading proponent of this view, but Lalande[23] lists many others before and after Kant: Plato, Descartes, Malebranche, Leibniz, Hegel, and closer to Durkheim, Ribot, Renouvier, Espinas, Herbart, Wundt, Hamelin, Taine, Fechner, Bergson and Guyau – though this list is not meant to be exhaustive. Note that Wundt and Herbart were pioneers of what has come to be known as contemporary psychology. But their focus on collective *and* individual representations is applicable to the broad range of the social sciences. Durkheim had studied under Wundt, who also influenced Simmel and Freud. In his book on 'totality,' Mauss[24] mentions almost the same list that Lalande invokes in this regard.

As for the concept of the 'will,' there was William James' 'Will to Believe' and Nietzsche's 'Will to Power,' which both Durkheim and Simmel specifically mention. Durkheim[25] depicts Tönnies' classic on community and society as a blend of Schopenhauer and Marx – witness Tönnies' opposition between the 'natural will' of community versus the 'rational will' of society.[26] Simmel transformed Schopenhauer's concept of 'will' into the concept of 'life,' that restless, powerful, irrational force that breaks through life's 'forms' and creates new ones.[27] Freud transformed it into the 'id' and, in the words of Ernest Jones, into 'the importance of the 'unconscious will''[28] as well as other representations of desire. Wundt referred to the conflict between the 'social will' and the 'individual will'.[29] Even Darwin referred to the 'will' in his evolutionary scheme.[30] Thus, there is no need to be dogmatic about the choice of Schopenhauer in this analysis. Schopenhauer was representative of his times, and for a time, he was immensely influential. We have by no means exhausted all the varieties of 'will to' something or other found in turn of the century writings.

Schopenhauer begins his famous *The World as Will and Idea* with the bold claim that 'The world is my idea.' This move was his way of reconciling the object-subject distinction that Kant had exposed with great force. Its consequence is that the objective world can never be known as a thing-in-itself but only in relation to a knowing subject, and that this knowledge is never fully 'objective' only an image. However, for Schopenhauer, the 'other side' of the representation, the thing-in-itself, is the 'will.'[31] It is important to note that Schopenhauer departed from Kant primarily in treating the 'idea' and 'will' as a conjunction, a unity.[32] Nevertheless, though will and idea are a unity, they are antagonistic to each other. In Schopenhauer's words, the will stands for the 'heart' and the idea for the mind.[33] The will encompasses dreams, impulses, affection, passions, and all that is obscure, unconscious and emotional.[34] Like Durkheim and Freud, Schopenhauer associates the will with 'the body' and its 'appetites,' 'passions,' and 'desires.' The mind stands for reflection, thought, abstraction, and the impersonal – all that controls or discounts the 'will.' But according to Schopenhauer, the heart is stronger than the mind.[35] The will is stronger than the representation. Schopenhauer reversed the Enlightenment understanding that human passion was an instrument of reason. For Schopenhauer, reason is the instrument of the blind, all-powerful will.[36] And, the tyrannical will is never satisfied:

All willing arises from want, therefore from deficiency, and therefore from suffering. The satisfaction of a wish ends it; yet for one wish that is satisfied there remain at least ten which are denied. Further, the desire lasts long, the demands are infinite; the satisfaction is short and scantily measured out. But even the final satisfaction is itself only apparent; every satisfied wish at once makes room for a new one; both are illusions; the one is known to be so, the other not yet. No attained object of desire can give lasting satisfaction, but merely a fleeting gratification.[37]

Humans suffer because the will is infinite; the more one has, the more one wants. The will 'is a striving without aim or end'[38] so that 'suffering is essential to life.'[34] It is easy to spot Durkheim's portrait of anomie as the 'bottomless abyss'[40] of human passions and Freud's tyrannical 'id'[41] as refractions of this aspect of Schopenhauer's thought. Human egoism is the cause of all social strife, according to Schopenhauer, and leads to immorality and unhappiness. Philosophers have justifiably labelled Schopenhauer the philosopher of pessimism.

But the late nineteenth century may also be called the age of pessimism,[42] and the founding fathers of the social sciences may justifiably be called pessimists. This point is passé for Freud, but consider Weber's gloomy depiction of the 'iron cage'[43] of civilization and Durkheim's gloomier still pronouncement on civilization, that suffering will increase as civilization progresses.[44] Simmel portrayed 'life' and its 'forms' as locked in a never-ending battle that the imperious 'life' will always win.[45] Tönnies[46] depicted 'society' itself as the result of the disintegration of 'community' which unleashed the 'will' from its perviously restrained state. Stress, conflict, tension, malaise and misery – these and related Schopenhauerian themes are found in the works of many late nineteenth century writers in diverse fields.

It must be emphasized that contrary to many contemporary accounts, Durkheim dethrones Comte as the founder of positivism or sociology in his *Socialism and Saint-Simon*.[47] Mauss ignores Comte. Durkheim's follower, Lucien Levy-Bruhl,[48] mocks Comte's own, unacknowledged metaphysics *despite* Comte's apparent contempt for metaphysics. Bouglé makes it clear that he cannot accept Comte's excessive rationalism.[49] The reason for this hostility toward Comte seems to be based on Comte's failure, relative to Schopenhauer, to account for the tyranny of human desires. In general, it seems that the alleged influence of Comte on Durkheimian sociology is a contemporary invention to justify the modern optimism that 'objective' studies will solve social problems. While it is possible to argue for Comte's influence in the sense that it is possible to argue anything, it is more difficult to justify using Comte's philosophy as the starting point for apprehending Durkheim given Durkheim's antipathy toward him and given the popularity of Schopenhauer's philosophy in Durkheim's time.

Schopenhauer's dualism undercuts the object-subject debate and the division of quantitative versus qualitative methodologies. Applying 'idea' and 'will' to both object and subject, successively, one can conclude the following: 'Objective facts' are not the things-in-themselves and 'hard' truths they are often made out to be. Rather, they, too, are only representations, images of 'reality' subject to interpretation and change that are nevertheless a kind of reality. Similarly, the actor's 'subjective' point of view is subject to illusion because the human agent is never fully aware of all the representations he possesses or that are acting upon him, nor is he fully aware of the imperious will. To complicate matters, collective *and* individual representations have a 'will' of their own that act upon the human agent and his representations and will. It would be a mistake to characterize Schopenhauer's distinction between the representation and will as a mere extension of the object subject distinction, for this was the very distinction he sought to transcend.[50] Students of Schopenhauer are careful to point out that by his use of the term 'representation,' Schopenhauer did *not* mean for us to distinguish between the image and what it is an image of.[51] Rather, Schopenhauer posits a *homo duplex* within a *homo duplex*, which in turn can be subdivided again, but which is always a unity. The representation *is* a kind of reality.[53] Durkheim seems to have followed Schopenhauer's lead in this regard. For example, in *The Rules of Sociological Method* Durkheim insisted that he 'had expressly stated and reiterated in every way possible that social life was made up entirely of representations.'[53] These representations are not the representations *of* human agents, are not an epiphenomenon, Durkheim takes pains to argue in the rest of his book.[54] But the individual is[55] further divided into mind versus body, representations versus will.[56] And society is further divided into representations versus will. Mauss echoes Schopenhauer when he mentions specifically that the concept of totality implies a *homo duplex* within a *homo duplex*.[57] Many of the founding fathers of the social sciences incorporated Schopenhauer's complexity, and this is what makes them fascinating as well as difficult to incorporate into positivistic methodologies.

In the methodologies of some of these precursors of the social sciences, the agent, witnesses and the scientist, who is simply a more refined kind of witness, are confronted by representations that run the span from individual to collective, with many hybrid varieties of representations that fall between these extremes. The agent, witnesses and the scientist also perceive each other as representations. Some of these representations coalesce, group and gain enough structure to become institutionalized. These institutionalized representations are identifiable by signs of habit for William James,[58] constraint for Durkheim, compulsion for Freud, reification for Marx, and as 'forms' for Simmel. Other representations tend to be more private, subjective and idiosyncratic. All of these representations are like little bundles of consciousness that combine with other bundles of consciousness. Thus, there will be *many* centers of consciousness and memories, even collective memories in society, beginning with and *within* the individual and developing through dyads, friendships, families, occupational groups on to 'society,' which in turn is composed of other smaller 'societies.' These writers, including Freud, deny any one center of consciousness, within or outside the agent. An excellent illustration is the Durkheimian follower Maurice Halbwachs' *The Collective Memory*,[59] which was an effort to synthesize Freudian and Durkheimian, as well as the social, psychological and physiological dimensions of the phenomena 'memory' and 'consciousness.'

Many of these representations, these bundles of consciousness, take on a life and 'will' of their own, and attach themselves synthetically to other representations on their own accord, quite apart from the human agent and the social structure. Thus, Durkheim's explanation of how the representation of the 'sacred' attaches itself to phenomena as diverse as totems, blood, property and the individual. Durkheim phrased the general principle as follows:

One cannot repeat too often that everything which is social consists of representations, and therefore is a result of representations. However, this act of becoming a collective image, which is the very essence of sociology, does not consist of a progressive realization of certain fundamental ideas. These fundamental ideas, at first obscured and veiled by adventitious beliefs, gradually liberate themselves and become more and more completely independent.[40]

These representations are not mere reflectors of reality. Durkheim explains:

A representation is not simply a mere image of reality, an inert shadow projected by things upon us, but it is a force which raises around itself a turbulence of *organic* and *psychical* phenomena.[61]

One important dimension is still missing in this complicated state of affairs – the concept of the unconscious.[62] The unconscious is a refraction of Schopenhauer's will, and Freud admitted this. But Durkheim also held an elaborate concept of the unconscious and constantly distinguished between society 'as it is' versus 'as it appears to itself,' and made similar distinctions on the level of the individual human agent's perceptions of private psychological states. Some portion of collective and individual representational life is always hidden from the apprehension of agents, witnesses, the scientist and society itself. All findings, quantitative or qualitative, are only representations. And the 'will' is always lurking in the shadows, waiting to destroy these seemingly objective and rational findings.

It follows that many of the precursors of the contemporary social sciences – especially Durkheim and Freud – did *not* follow the plan of theory, hypothesis testing, empirical testing, and so on that are supposed to be the staple of positivistic research. Rather, some of them understood all comparative-historical research as well as micro-research as aspects of the one 'well-designed experiment.' Schopenhauer advocated this position, reasoning that the nature of phenomena objectifies itself in one case as well as in countless cases.[63] Facts and findings *cannot* be 'verified' in their scheme of things because facts and findings are representations. Each time facts are approached, the object, human agent doing the analysis, and witnesses 'observing' the interaction between object and subject have changed. Each fact is relative to the configuration of object, subject and witnesses. Human will interacts with 'objective' reality because the scientist dictates which questions will be asked of 'reality' and how they will be asked. Hypotheses are not discarded just because they cannot be verified repeatedly. Rather, each experiment yields knowledge which is useful in some fashion. In a sense, scientific discourse strings out in time the multiple meanings contained in a single moment of ordinary or artistic discourse. Both art and science deal with representations, not 'reality,' and both involve the temporary suspension of the 'will.'[64] Just as there can be no last and lasting interpretation of any artistic work, so also any scientific description can be superceded at any time.

DURKHEIM'S *SUICIDE* AS A STUDY IN TOTALITY

The stamp of Schopenhauer on Durkheim's *Suicide* is immediately apparent even though Durkheim does not refer to Schopenhauer by name in this work. If one pays attention to Schopenhauer's doctrine on perception, the will, the antagonistic dualism of human nature, pessimism, and condemnation of egoism as the basis of immorality,[65] one will find all of these reproduced in Durkheim's *Suicide*. The most compelling evidence for Schopenhauer's influence upon Durkheim lies in Durkheim's obsession with pain and suffering throughout this argument, and his linkage of that torment to the inherent insatiability of human desires. Humans suffer because of the infinite will which must be contained by the representational system that is society.[66] Ironically, this is precisely the aspect of Durkheim's thought that many contemporary functionalists have tried to modify. Consider, for example, Merton's opening arguments in his famous essay 'Social Structure and Anomie.'[67] Merton claims that he wants to move away from the Freudian-like perspective in which social disorder is viewed as the result of the ungoverned id and other biological tendencies. But to the extent that Merton ascribes his conceptualization of anomie to Durkheim – which he does at times – one must account for the fact that Durkheim's concept of the dualism of human nature has many affinities with the Freudian view – itself a modification of Schopenhauer – that Merton rejects. Maurice Halbwachs exposes Schopenhauer's influence when he writes, commenting on Durkheim's *Suicide*, that

We can assume that the number of suicides is a rather exact indicator of the amount of suffering, malaise, disequilibrium, and sadness which exists or is produced in a group. Its increase is the sign that the sum total of despair, anguish, regret, humiliation, and discontent of every order is multiplying.[68]

Suicide is a book about socio-psycho-physical pain in various forms. The seemingly obvious role of pain in suicide has been missed in many commentaries on Durkheim's work. Alvarez has a point when he charges that despite all that sociologists have written about suicide, they have made the subject seem somehow 'unreal.'[69] For example, in the functionalist version, anomie has no feel to it – yet Durkheim describes the anguish, torment and suffering associated with anomie with incredible pathos. Ironically, the Parsonian 'war of all against all' is quite bland. Moreover, note that Halbwachs seems to imply an intricate relationship between individual and group suffering. We shall be tracing some of that intricacy in this section.

In the opening moves of *Suicide*, Durkheim claims that he 'shall determine the nature of the social causes [of suicide, and] how they produce their effects, *and* their relations to the individual states associated with the different sorts of suicide.'[70] Clearly, he seems to be aiming at total explanation from the start. Perhaps his aim would have been better understood had he launched next into a discussion that comes toward the end of the book. Durkheim takes up the question whether explanations of suicide on the level of what happens to individuals are adequate. In many ways, he foreshadows the most dominant line of research today, what has come to be known as the 'stressful life events school.'[71] Durkheim writes:

There is however another hypothesis, apparently different from the above, which might be tempting to some minds. To solve the difficulty, might we not suppose that the various incidents of private life considered to be preeminently the causes determining suicide, regularly recur annually in the same proportions? Let us suppose that every year there are roughly the same number of unhappy marriages, bankruptcies, disappointed ambitions, cases of poverty etc. Numerically the same and analogously situated, individuals would then naturally form the resolve [to commit suicide] suggested by their situation, in the same numbers. One need not assume that they yield to a superior influence; but merely that they reason generally in the same way when confronted by the same circumstances.[72]

It could be said that Durkheim is summarizing and synthesizing the contemporary version of the understanding of society as the 'outcome of human agency'[73] and the 'subjective points of view of individuals.'[74] In much of the literature on stress, the researchers try to show that exposure to stressful life events can lead individuals to a host of symptoms, including suicide. Durkheim's aim is not so much to discount something like this subjectivist argument as to reconcile it with objectivism.

Essentially, Durkheim wants to know why these 'stressful life events' recur regularly in a given society. He seems to imply that social rates of birth, death, marriage, divorce and many other items on lists of stressful life events are 'objective' in the sense that they are characteristic of the group of which the individual forms a part. Furthermore, Durkheim wonders why a certain number of individuals are regularly predisposed to succumb to these events. In other words, there are social rates of causes of human suffering *and* social rates of individual predisposition to that suffering. Durkheim forces one to confront the extreme objectivist *and* subjectivist versions of the stressful life events argument. The effects of stressful life events are shaped by collective processes *and* individual predisposition. Thus, he writes:

But we know that these individual events, though preceding suicides with fair regularity, are not their real causes. To repeat, no unhappiness in life necessarily causes a man to kill himself unless he is otherwise so inclined. The regularity of possible recurrence of these various circumstances thus cannot explain the regularity of suicide. Whatever influence is ascribed to them, moreover, such a solution would at best change the problem without solving it. For it remains to be understood why these desperate situations are identically repeated annually, pursuant to a law peculiar to each country. How does it happen that a given, supposedly stable society always has the same number of disunited families, of economic catastrophies, etc.? This regular recurrence of identical events in proportions constant within the same population but very inconstant from one population to another would be inexplicable had not each society definite currents impelling its inhabitants with a definite force to commercial and industrial ventures, to behaviour of every sort likely to involve families in trouble, etc. This is to return under a very slightly different form to the same hypothesis which had been thought refuted.[75]

Durkheim's reasoning on this issue is compelling. Societies definitely exhibit characteristic rates of many 'stressful life events.' And societies just as regularly produce certain rates of depression, alcoholism and other forms of mental illness that predispose individuals to succumb to life's sorrows. Thus, Durkheim is clearly trying to conjoin society *and* the individual in his argument.

It should be noted that in contrast to Durkheim, the contemporary stressful life events school of research does not emphasize the social rates of stressful events.[76] The unit of analysis, in general, is the individual. Individuals are said to experience certain events and some of these individuals fall ill. Both extremes of *homo duplex* are missing from this explanation. One is society's role in the genesis of the events, and the other is society's role in the individual's predisposition to succumb to illness.

Consider Durkheim's scattered discussion of 'neurasthenia,' which may be roughly translated into the contemporary 'depression,' in the context of his quest for a 'total' explanation. According to Durkheim, neurasthenia predisposes the individual to suicide because, on the psychological *and* physiological levels of explanation, such an individual is overly sensitive to pain and consequently suffers too much. The pain drives the individual to abandon life and weakens his instinct of self-preservation in the manner suggested by Mauss. Durkheim describes the neurasthenic as delicate, sensitive, easily stimulated, unstable, and 'destined to suffer.'[77] Translated into the more modern vocabulary of psychologists like Asenath Petrie, one could characterize Durkheim's neurasthenics as 'augmenters' as opposed to 'reducers.'[78] Augmenters suffer because they are psychologically and physically overly sensitive to stimuli. According to Durkheim, 'this psychological type [neurasthenia] is therefore very probably the one most commonly to be found among suicides.'[79] But Durkheim shows that neurasthenia is not, by itself, a cause of suicide. For example, women suffer from neurasthenia much more than men, but men have higher suicide rates than women. According to Durkheim, 'if there were a causal relation between the suicide-rate and neurasthenia, women should kill themselves more often than men.'[80] Clearly, some other factor is at work.

Durkheim regards neurasthenia as the prototype for all kinds of madness. If neurasthenia or mental illness in general predispose the individual to commit suicide, suicide rates and rates of mental illness should be positively related. But, according to Durkheim, 'the countries with the fewest insane have the most suicides.'[81] In addition, the same social forces that produce the social causes of suicide also tend to produce neurasthenia: 'The hypercivilization which breeds the anomic tendency and the egoistic tendency also refines nervous systems, making them excessively delicate; through this very fact they are less capable of firm attachment to a definite object, more impatient of any sort of discipline, more accessible both to violent irritation and to exaggerated depression.'[82] Hence,

A given number of suicides is not found annually in a social group just because it contains a given number of neuropathic persons. Neuropathic conditions only cause the suicides to succumb with greater readiness to the current. Whence comes the great difference between the clinician's point of view and the sociologist's.[83]

Thus, in Durkheim's view, the individual is predisposed to suicide based upon his own subjective hypersensitivity to pain, and part of that predisposition is psychological and physiological. But society also plays a role in that subjective predisposition along the lines of the 'total social fact.'

Durkheim's treatment of the social causes of suicide can be divided into two distinct types. Both are sensitive to the role of the individual, but in different ways. One is concerned with the effects of exposure to society on suicide illustrated through the suicide rates of females relative to males, children relative to the aged, daytime versus the night, the spring and summer versus fall and winter and other indicators of 'heightened social life.' This approach corresponds roughly to present-day quantitative approaches, though, given Durkheim's philosophical premises, it does not correspond exactly. The other, more qualitative approach, addresses the issue of society's integration and disintegration, and the effects of these societal states, as filtered through the aforementioned degree of social exposure, upon individuals. It is important to keep these two levels of argument radically distinct, and to note the differential role of the 'will' in each.

For example, the most common misinterpretation of the gist of Durkheim's argument in *Suicide* is that suicide is inversely related to social integration such that integration is misconceived as the attachment of individuals to groups.[84] Closely related is the corollary that social supports, contacts and ties should act as a prophylactic against suicide.[85] These misunderstandings are a distortion of Durkheim's statement on integration and are also a conjunction of two separate arguments, mentioned above. Durkheim did *not* refer to social integration as a property of individual attachment *to* groups, but as a property *of groups*: 'So we reach the general conclusion: suicide varies inversely with the degree of integration of the social groups of which the individual forms a part.'[86] Moreover, even if one assumes for the sake of argument that Durkheim did intend integration to refer to social ties, one encounters difficulties. Contemporary sociologists focus on the proposition that suicide varies inversely with integration, and ignore his claims that it can also be caused by excessive integration, as in altruistic suicide.[87] Suppose that one drops the term 'integration' from such discussions because of this problem. That still leaves the fact that Durkheim argued that other versions of social ties, contacts and supports can be pathogenic.

In fact, Durkheim argues that social contacts in disintegrating societies are pathogenic. Note that we have phrased this statement such that the two arguments mentioned above are kept distinct: society's degree of integration is a phenomenon different and distinct from exposure to society. Durkheim repeatedly argues against the most dominant interpretation of his thought, the idea that social contacts are beneficial. For example, he showed that women were shielded from suicide precisely because they participated less than men in society, that the aged have higher suicide rates than children because the aged have had longer exposure to society, and that suicide rates increase with the lengthening of the day because longer days enable greater exposure to social interaction. Thus, with regard to women and social contacts, Durkheim writes:

If women kill themselves much less often than men, it is because they are much less involved than men in collective existence; thus they feel its influence – good or evil – less strongly.[88]

Similarly, 'day favors suicide because this is the time of most active existence, when human relations cross and recross, when social life is most intense.'[89] Spring and Summer months favor suicide because they coincide with the lengthening of the day.[90] Suicide 'reaches its height only in old age'[91] because society has had more time to 'penetrate' the consciousness of the old person while 'society is still lacking'[92] in the child. Durkheim goes so far as to claim that 'the immunity of an animal has the same causes,' namely, the impossibility of human society to penetrate too deeply into animal consciousness.[93]

Durkheim foreshadows current findings that suicide rates among children and teen-agers are increasing, but he offers a 'total' explanation.[94] Contemporary attempts at explaining youth suicide, among laypersons and professionals alike, assume that the suicide of young people is a unique phenomenon relative to suicide in general.[95] Durkheim implies that any person, young or old, is more likely to kill himself if he is overexposed to society. It should be kept in mind that little boys kill themselves more than little girls and that in general, the suicide patterns of the young reflect overall suicide patterns: suicide among the young is still a predominantly Protestant, urban, affluent 'hyper-civilized' phenomenon. Durkheim writes:

It must be remembered that the child too is influenced by social causes which may drive him to suicide. Even in this case their influence appears in the variations of child-suicide according to social environment. They are most numerous in large cities. *Nowhere else does social life commence so early for the child, as is shown by the precocity of the little city-dweller.* Introduced earlier and more completely than others to the current of civilization, he undergoes its effects more completely and earlier. This also causes the number of child-suicides to grow with pitiful regularity in civilized lands.[96]

Essentially, Durkheim is claiming that young people in modern societies 'grow up' too fast. Their 'will' is subjected to the 'infinity of desires' earlier than it used to be in traditional societies. But in this as in other cases, social contacts, *not* some alleged *lack* of social attachment to groups, contribute to suicide. If anything, social attachment is the culprit. Durkheim's assessment is relevant to contemporary discussions of youth suicides, but it is not typically invoked in discussions of this sort.

Why would exposure to society be pathogenic? Because society itself is *disintegrating* in most Western societies, and this societal state unleashes the will, which produces suffering directly as well as a predisposition to suffer. Thus, keeping the two levels of argument discussed above distinct, but combining them, one can say that the individual's exposure to a disintegrating society is harmful to the individual. Lack of social contacts in such a situation may actually be beneficial. According to Durkheim,

Individuals share too deeply in the life of society for it to be diseased without their suffering infection. What it suffers they necessarily suffer. Because it is the whole, its ills are communicated to its parts. Hence it cannot disintegrate without awareness that the regular conditions of general existence are equally disturbed ... Thence are formed currents of depression and disillusionment emanating from no particular individual but expressing society's state of disintegration.[97]

New metaphysical and religious systems arise which reflect this 'collective asthenia, or social malaise' and 'they merely symbolize in abstract language and systematic form the physiological distress of the body social.'[98] Note that the individual and society are portrayed by Durkheim as existing in a very intimate relationship. This fact also attests to his efforts to reconcile object and subject in his explanations.

Egoism and anomie, which Durkheim claims are conjoined, are the result of society's disintegration and 'disaggregation'[99] such that 'the individual ego asserts itself to excess in the face of the social ego and at its expense.'[100] Using Schopenhauer's vocabulary, one could summarize Durkheim's claims to the effect that society's disintegration causes the unleashing of the insatiable 'will.' Durkheim follows Schopenhauer's lead in claiming that 'our capacity for feeling is in itself an insatiable and bottomless abyss.'[101] Thus, religion, marriage, and political institutions all regulate 'the life of passion.'[102] But in a disintegrating, disaggregated society characterized by a state Durkheim calls *dérèglement* (meaning derangement), the will and its manifestations – the passions, unbridled aspirations, the longing for infiniteness – reign. We can infer from Durkheim's other writings that society is not integrated when it is structured such that the 'lower' aspect of *homo duplex* – which has to do with the will – is allowed to rule over the 'higher.'[103] For example, Durkheim accuses both capitalism and socialism of promoting anomie and egoism because they assume that human desires will regulate themselves of their own accord.[104] According to Durkheim, this is a mistake. He implies Schopenhauer's assessment that the 'will' is inherently unruly and can never serve as the basis for human morality or contentment.

Undoubtedly, societal disaggregation causes the individual's ties to society to become slack. Durkheim makes claims to this effect often enough, most frequently in *Moral Education*.[105] But that fact should not make one lose sight of the more important point that integration and disintegration refer to *qualities* of the group, which in turn affect the individual's attachment to the group. Durkheim conjoins qualitative and quantitative approaches especially with regard to the concept of 'integration.' To misunderstand this point is to become entangled in the 'over-socialized view of man'[106] which misrepresents Durkheim's position. Durkheim is not arguing that social ties are always beneficial, as if society were the penultimate reality and the individual counted for nothing. Rather, he reproduces Schopenhauer's opposition between the 'idea' and the tyrannical 'will' in a new form. For Durkheim, 'essentially social life is made up of representations'[107] – society is a system of ideas. On the other side, man's imperious 'desires,' 'appetites' and 'passions' – various versions of Schopenhauer's 'will' – are always working in opposition to society as idea. To repeat, the 'will' is stronger than the idea, desire is all-powerful. When society disintegrates, its representations of itself become pessimistic and pathogenic, and it is no longer in a position to act as a check on the 'will.' In this state of affairs, lack of attachment to the disintegrating group is beneficial at the same time that it holds the potential for further unravelling of the 'will.' The final outcome depends on the quality *and* extent of exposure to society.

In sum, if one wants to be faithful to Durkheim's complex commitment to 'totality,' it is impossible to pick one strand of that thought without invoking the entire edifice, or one pole of any epistemological dualism without invoking the other. One pole of *homo duplex*, in any form, immediately implies the opposite pole. Start with the individual and his subjective point of view, and you must confront society's role in that subjective predisposition. If you begin with social integration, you must account for the power of individual 'will' which is working to undermine it. The quantitative sounding analysis of the degree of exposure to society – social contacts – is offset by the more qualitative analysis of the nature of that exposure – social bonds. The seemingly qualitative predisposition to suffering can be quantified in Durkheim's scheme. The seemingly quantitative analyses of divorce rates and other social rates are also, in part, representations of the experience of subjective suffering. And so on. Durkheim's argument in *Suicide* is much more intricate than frequently supposed, and it bridges the distinction between quantitative and qualitative methodology at almost every turn. Schopenhauer's insistence that the idea and will are a unity explains many of Durkheim's theoretical moves that positivistic readings have found baffling.

ACKNOWLEDGEMENTS

The research for this essay was supported, in part, by a Fellowship from the National Endowment for the Humanities in 1987, for which I am very grateful.

NOTES

[1] Robert Bogdan and Steven J. Taylor, *Introduction to Qualitative Research Methods*, New York, John Wiley & Sons, 1975, pp. 1–21.

[2] For recent examples, see Anthony Flew, *Thinking About Social Thinking: The Philosophy of the Social Sciences*, New York, Basil Blackwell, 1986; David Sylvan and Barry Glassner, *A Rationalist Methodology for the Social Sciences*, New York, Basil Blackwell, 1986; Roger Trigg, *Understanding Social Science*, New York, Basil Blackwell, 1985; Jerry A. Fodor, *Representations: Philosophical Essays on the Foundations of Cognitive Science*, Cambridge, MIT Press, 1984.

[3] Irving L. Horowitz, 'Disenthralling sociology,' *Society* 24, 1987, pp. 48–55.

[4] Marcel Mauss, *Sociology and Psychology*, London, Routledge & Kegan Paul, [1950] 1979.

[5] *Ibid.*, p. 24.

[6] *Ibid.*

[7] *Ibid.*, p. 25

[8] Célestin Bouglé, *The French Conception of 'Culture Générale' and Its Influences upon Instruction*, New York, Columbia University Press, 1938, p. 24.

[9] *Ibid.*, p. 22. See also Stjepan G. Meštrović, 'Durkheim's Concept of Anomie Considered as a 'Total' Social Fact,' *British Journal of Sociology* 38 (4), 1987, pp. 567–583.

[10] Arthur Schopenhauer, *The World as Will and Idea*, New York, AMS Press, [1818] 1977.

[11] Bouglé, *op. cit.*, p. 21.

[12] My interview was conducted in Paris in 1987 with Pierre Halbwachs, retired professor of linguistics at the Sorbonne in Paris, and son of Maurice Halbwachs, who was one of Durkheim's most prolific followers.

[13] André Lalande, 'Allocution pour le centenaire de la naissance d'Emile Durkheim,' *Annales de l'Université de Paris*, 1960, p. 23.

[14] See Steven Lukes, *Emile Durkheim: His Life and Work*, New York, Harper & Row, 1972, p. 21, though Lukes does not invoke Schopenhauer. See also Stjepan G. Meštrović, 'Durkheim's conceptualization of political anomie,' *Research in Political Sociology*, 1988, forthcoming and Stjepan G. Meštrović, 'Durkheim, Schopenhauer

and the Relationship Between Goals and Means: Reversing the Assumptions in the Parsonian Theory of Rational Action,' *Sociological Inquiry*, 1988, forthcoming.

[15] Sigmund Freud, *The Standard Edition of the Complete Psychological Works of Sigmund Freud*, Vol. 20, 1974, pp. 59–60. Henri Ellenberger, *The Discovery of the Unconscious*, New York, Basic Books, 1970, treats Schopenhauer's influence on Freud at some length. See also Ernest Jones, *The Life and Work of Sigmund Freud*, New York, Basic Books, 1981, Vol. 1, pp. 319, 375; Vol. 2, pp. 226, 415; Vol. 3, pp. 205, 313.

[16] André Lalande, *Vocabulaire technique et critique de la philosophie*, Paris, Presses Universitaires de France, [1926] 1980, p. 1221.

[17] Bryan Magree, *The Philosophy of Schopenhauer*, New York, Oxford University Press, 1983, p. 264 and pp. 379–390.

[18] David W. Hamlyn, *Schopenhauer*, London, Routledge & Kegan Paul, 1980; Patrick Goodwin, 'Schopenhauer,' *The Encyclopedia of Philosophy*, Vol. 8, New York, Macmillan, 1967, pp. 325–32.

[19] Thomas Mann, 'Introduction,' in W. Durant (ed.), *The Works of Schopenhauer*, New York, Frederick Unger Publishers, [1939] 1955, pp. iii–xxiii.

[20] Georg Simmel, *Schopenhauer and Nietzsche*, Amherst, University of Massachusetts Press, [1907] 1986. See also Stjepan G. Meštrović, 'Simmel's concept of the unconscious,' paper presented to the Western Social Science Association in San Diego, 1984.

[21] Stjepan G. Meštrović, 'Durkheim's renovated rationalism and the idea that 'collective life is only made of representations',' *Current Perspectives in Social Theory* 6, 1985, pp. 199–218.

[22] Alan Janik and Stephen Toulmin, *Wittgenstein's Vienna*, New York, Simon & Schuster, 1973.

[23] André Lalande, [1926] 1980, *op. cit.*, pp. 920–22.

[24] Marcel Mauss, *op. cit.*

[25] Emile Durkheim, 'Review of Ferdinand Tönnies, Gemeinschaft und Gesellschaft,' in Mark Traugott, ed., *Emile Durkheim on Institutional Analysis*, Chicago, University of Chicago Press, 1978, pp. 115–122.

[26] Ferdinand Tönnies, *Community and Society*, New York, Harper & Row, [1887] 1963.

[27] Georg Simmel, *'Georg Simmel on Individuality and Its Social Forms*, Chicago, University of Chicago Press, 1971. See especially pp. 376–91.

[28] Ernest Jones, *op. cit.*, Vol. 2, p. 226. See also Magee, *op. cit.*, p. 283.

[29] Wilhelm Wundt, *Ethics*, New York, Macmillan, [1886] 1902, pp. 58–96.

[30] Bryan Magee, *op. cit.*

[31] Schopenhauer, *op. cit.*, p. 124.

[32] *Ibid.*, p. 32.

[33] *Ibid.*, Vol. 3, pp. 105–118.

[34] *Ibid.*, Vol. 1, pp. 405–20.

[35] *Ibid.*, Vol. 3, p. 79.

[36] *Ibid.*, p. 77.

[37] *Ibid.*, Vol. 1, p. 254.

[38] *Ibid.*, p. 414.

[39] *Ibid.*, p. 410.

[40] Emile Durkheim, *Suicide*, New York, Free Press, [1897] 1951.
[41] Sigmund Freud, 'The ego and the id,' in *The Standard Edition, op. cit.*, Vol. 19, pp. 1–59.
[42] Thomas Mann, *op. cit.*
[43] Max Weber, *The Protestant Ethic and the Spirit of Capitalism*, New York, Charles Scribner's Sons, [1904–1905] 1958.
[44] Emile Durkheim, 'The dualism of human nature and its social conditions,' in R. Bellah (ed.), *Emile Durkheim on Morality and Society*, Chicago, University of Chicago Press, [1914] 1973, pp. 149–66.
[45] Georg Simmel, *op. cit.*, 1971.
[46] Ferdinand Tönnies, *op. cit.*
[47] Emile Durkheim, *Socialism and Saint-Simon*, Yellow Springs, Antioch Press, [1928] 1958.
[48] Lucien Levy-Bruhl, *The History of Philosophy in France*, Chicago, Open Court, 1899.
[49] Bouglé, *op. cit.*
[50] Hamlyn, *op. cit.*, p. 45; Magee, *op. cit.*, p. 35.
[51] Hamlyn, *op. cit.*, p. 5.
[52] *Ibid.*, p. 45, and Magee, *op. cit.*, p. 76.
[53] Emile Durkheim, *The Rules of Sociological Method*, New York, Free Press, [1895] 1983, p. 34.
[54] Emile Durkheim, *Sociology and Philosophy*, New York, Free Press, [1924] 1974.
[55] Durkheim, *op. cit.*, [1914] 1973.
[56] Stjepan G. Meštrović, *In the Shadow of Plato: Durkheim and Freud on Suicide and Society*, doctoral dissertation, Syracuse University, 1982.
[57] Mauss, *op. cit.*
[58] Charles Camic, 'The matter of habit,' *American Journal of Sociology* 91, 1986, pp. 1039–87.
[59] Maurice Halbwachs, *The Collective Memory*, New York, Harper & Row, [1950] 1980.
[60] Emile Durkheim, *Incest: The Nature and Origin of the Taboo*, New York, Lyle Stuart, [1897] 1963, p. 114.
[61] Emile Durkheim, *The Division of Labor in Society*, New York, Free Press, [1893] 1933, p. 97, emphasis added.
[62] Stjepan G. Meštrović, 'Durkheim's concept of the unconscious,' *Current Perspectives in Social Theory* 5, 1984, pp. 267–88.
[63] Hamlyn, *op. cit.*, pp. 95–105.
[64] See Schopenhauer's scattered discussion of art as a state of 'will-lessness.'
[65] Goodwin, *op. cit.*
[66] Sjepan G. Meštrović and Helene M. Brown, 'Durkheim's concept of anomie as dérèglement,' *Social Problems* 33, 1985, pp. 81–99.
[67] Robert K. Merton, *Social Theory and Social Structure*, New York, Free Press, 1957.
[68] Maurice Halbwachs, *The Causes of Suicide*, [1930] 1978, London, Routledge & Kegan Paul, p. 314.
[69] A. Alvarez, *The Savage God*, New York, Bantam, 1970.
[70] Durkheim, *op. cit.*, [1897] 1951, p. 52, emphasis added.

118   STJEPAN G. MEŠTROVIĆ

71 For discussions in the context of Emile Durkheim, see Stjepan G. Meštrović and Barry Glassner, 'A Durkheimian hypothesis on stress,' *Social Science and Medicine* 17, 1983, pp. 1315–27 and Stjepan G. Meštrović, 'A sociological conceptualization of trauma,' *Social Science and Medicine* 21, 1985, pp. 835–48.
72 Durkheim, *op. cit.*, [1897] 1951, pp. 305–6.
73 Anthony Giddens, *The New Rules of Sociological Method*, New York, Basic Books, 1976.
74 Bogdan and Taylor, *op. cit.*
75 Durkheim, *op. cit.*, [1897] 1951, p. 306.
76 Meštrović, *op. cit.*, 1982, pp. 239–75.
77 Durkheim, *op. cit.*, [1897] 1951, p. 68.
78 Asenath Petrie, *Individuality in Pain and Suffering*, Chicago, University of Chicago Press, 1967.
79 Durkheim, *op. cit.*, [1897] 1951, p. 69.
80 *Ibid.*, p. 71.
81 *Ibid.*, p. 73.
82 *Ibid.*, p. 323.
83 *Ibid.*
84 For reviews of the use of the concept of 'integration,' see Meštrović, *op. cit.*, 1982 and Meštrović & Glassner, *op. cit.*, 1983.
85 *Ibid.*
86 Durkheim, *op. cit.*, [1897] 1951, p. 209.
87 *Ibid.*, p. 221.
88 *Ibid.*, p. 299.
89 *Ibid.*, p. 117.
90 *Ibid.*, p. 119.
91 *Ibid.*, p. 101.
92 *Ibid.*, p. 215.
93 *Ibid.*
94 For a recent example, see G. Klerman (ed.), *Suicide and Depression Among Adolescents and Young Adults*, Washington, D.C., American Psychiatric Association, 1986.
95 *Ibid.*
96 Durkheim, *op. cit.*, [1897] 1951, pp. 100–1, emphasis added.
97 *Ibid.*, p. 214.
98 *Ibid.*
99 *Ibid.*, pp. 221, 289.
100 *Ibid.*, p. 209.
101 *Ibid.*, p. 247.
102 *Ibid.*, p. 270.
103 See Meštrović, *op. cit.*, 1982.
104 Emile Durkheim, *Professional Ethics and Civic Morals*, Westport, Greenwood Press, [1950] 1983, pp. 1–17.
105 Emile Durkheim, *Moral Education*, Glencoe, Free Press, [1925] 1961.
106 Dennis Wrong, 'The over-socialized conception of man in modern sociology,' *American Sociological Review* 26, 1961, pp. 183–93.
107 Durkheim, *op. cit.*, [1897] 1951, p. 312.

ROBERT FELEPPA

CULTURAL KINDS:
IMPOSITION AND DISCOVERY IN ANTHROPOLOGY

Qualitative analysis in cultural anthropology is often viewed as burdened with the special requirement that it produce units of description, measurement, and comparison that embody the conceptions of the society under study. The nature and extent of this requirement varies, with some interpretive methodologies seeking to place subject conceptions among the most fundamental in the inquirer's analytic framework. Others make interpretation less methodologically pivotal, but it is generally rare that anthropologists will not mold the character of some units of analysis partly on the basis of translation of the source language.

Roughly, the reasons motivating this special constraint are these: (1) Anthropology's primary interest is in discerning how members of cultures conceptualize their social and natural world. (2) Inquirers typically have cultural and linguistic backgrounds that are alien to the people they study but which are all too easily attributable to them in place of their actual notions. Just as physical units must fit physical reality, social units must conform to a social reality: If, for instance, we are analyzing kinship and residence patterns (quantitatively, perhaps) we are to remember that anthropology's task is not simply to give a systematic accounting of those patterns from the inquirer's 'external' or 'etic' perspective, but to gain the subject's 'internal' or 'emic' one. Physicists can give us schemes of things, but social inquirers must give schemes *of* schemes of things. Thus C. H. Brown remarks:

> Systems of weights and measures, like all tools, are designed to meet certain requirements extraneous to their own internal logic. One would not, for example, weigh letters in fractions of tons, nor concrete blocks in multiples of ounces. As for measures, speed cannot be measured in meters nor a football field in miles per hour. I would be surprised to discover that kin terminologies in their capacity as linguistic tools do not similarly 'fit' the reality they are used to describe? (1974:429).

Thus we see in rough terms the motivation for discovering and perhaps employing in anthropological description expressions corresponding to what I here term the 'cultural kinds' of the people being studied. However, these concerns are double-edged. An important task of ethnography and ethnology is to communicate cultural facts to an audience whose idiom and background

119

Barry Glassner and Jonathan D. Moreno (eds.),
The Qualitative-Quantitative Distinction in the Social Sciences. 119–153.
© 1989 *by Kluwer Academic Publishers.*

is typically alien to the culture being described. Whatever culturally-specific concepts are utilized, the product must be significant to cultural outsiders. Analysis must conform to notions familiar to them, and if this entails 'imposing the familiar' in a way that compromises the description of subject categories then the basic aim of interpretation would seem unachievable. This is not to challenge the idea that a potentially valuable, 'expansion of the western mind' can result from anthropological study. For such novel conceptualization on the *inquirer's* part need bear no identity to what *subjects* conceive. However, if anthropology is supposedly a study *of* culture, it should be able to distinguish ethnographic fact from (albeit enlightening) social-science fiction.

Some see in this problem sufficient reason to give little or no weight at all to interpretation. Others see reason to revise or augment the interpretive methodologies so that objective interpretation is achieved. Most try to find some sort of hermeneutic compromise, a productive combination of etic and emic, but the nature and possibility of such a compromise remains unclear. One reason this is so is that, as is often the case with key terms in inter-paradigmatic controversy, the usage of 'emic' and 'etic' (much like 'quality' and 'quantity') varies considerably – at times to the degree that one person's 'emic' is another person's 'etic.' After sorting through some of these variations in order to identify terminological pitfalls for the unwary visitor, and in order to see the degree to which these variations signal deeper philosophical differences, I shall sketch in outline of a philosophical justification for translation-based inquiry (although I shall not insist that it is the only way to study culture). This justification will confront the pervasive problem of interpreter imposition, which, while seemingly threatening to interpretive (and thus emic-ethnographic) objectivity, can be seen in a more constructive light.[1]

EMICS/ETICS AND QUALITY/QUANTITY

Thus far I have used 'qualitative analysis' to apply simply to the development of conceptual frameworks for empirical inquiry. In its most systematic form it comprises the articulation of sets of definitionally interrelated terms from a small base of undefined primitives. So construed, qualitative analysis is hardly incompatible with quantitative – instead it is an important complement of the latter, as quantitative analysis will founder if units of measure are inadequate. And, on the face of it, use in quantitative analysis of emic units – say, whose labels have translational correlates in the source idiom – seems quite feasible, and has been employed, for instance, in Geoghegan's (1969) quantitative study of residence patterns.

However, this provides little by way of grasp of the source of any emic/etics or quality/quantity *controversy*. Here broader, though vaguer significations are relevant. Key features of quality/quantity questions are curtly summarized by Thomas Kuhn as follows:

On some occasions it has meant simply numerical measurement versus the direct apprehension of the senses. On other occasions, under the same rubric, the poverty of scientific abstractions (qualitative or quantitative) is opposed to the sensuous richness of life in the everyday world. At some points the opposition is between the positivist interpretation of the significance of scientific laws and the 'realist' view that such laws give us certain direct information about the 'true' structure of the universe. At still other points the opposition is between those who see the universe as an infinite continuum and those who see it as a finite discretum. This last formulation I find particularly puzzling, because in this case I am not even sure which party should be seen as representing quality and which quantity. Though all of these oppositions have been very real and all have had, at one time or another, historical relations to the others, I am at least dubious about the possibility of viewing them all as aspects of some single fundamental historical dichotomy. (Lerner 1961: 15f.)

These general themes resound in both the sociological and anthropological literatures, although 'quality/quantity' predominates in the former and 'emic/etic' in the latter. For instance, paralleling key elements of sociology's *Methodenstreit*, some interpretive anthropologists challenge the idea of approaching culture in terms of 'scientistic' strategies that aim to subsume a society's behavior to scientific law. In this connection one hears remarks reminiscent of Peter Winch's complaints that these strategies utterly miss the 'rule-governed' character of symbolic behavior and obscure conceptually organized action categories.[2] One also hears the related complaint that analytic strategies too infused with scientific abstraction distance inquirers from the social realities they study and in turn impoverish their understanding of them. Interpretive anthropologists complain that 'platonistic' formalism distorts more fluid and diverse cultural realities. This calls to mind parallel Frankfort School concerns that quantitative sociology's results are trivialized by over-tailoring to methodological rigor.

As with quality and quantity, emics/etics controversies are similarly irreducible to some single contrast, but similarly cluster about the aforementioned question of *fit* of descriptive theoretical apparatus with its object reality – a problem exacerbated by the cultural distance that often exists between inquirer and subject. This adds to epistemological, methodological, and terminological problems that have caused as much confusion and miscommunication as they have deeper and more substantial controversies. Not the least of the contributing causes is the fact that the terms 'etic' and 'emic' are used to characterize both dichotomous contrasts and complementary concepts.

THE ORIGINAL MODEL FROM PHONOLOGY

Complementarity is emphasized in Pike's derivation of the emics/etics distinction from the phonological notions of phonemics and phonetics (1954). Pike's idea was that if one could determine phonemes by systematically varying phonetic features of an expression so as to see what variations affected significance and which did not, one should similarly be able to identify defining criteria of expressions by elicitation procedures involving the variation in the properties of objects designated by those expressions. Varying, say, the height of someone properly called a bachelor will make no difference to their bachelorhood; while suitably varying their age, sex, or marital status will. By determining 'complementarity and contrast' we can tell what sounds make a difference to the subject ear: this is phonetic analysis. Measuring or describing the sound in terms of a universally applicable phonetic vocabulary, on the other hand, requires no characterization of the import of sound character or variation from the subject's point of view. Thus despite the complementarity, the phonological roots of the emic/etic contrast suggest a notion of emics as involving grasp of significance 'from the subject's point of view' vs. etics as grasping it 'from the inquirer's point of view.' There is also the suggestion of a contrast between emic as culturally specific or unique and as ascertainable by interpretive means as opposed to etic as universal and instrumentally measurable and quantifiable.

Pike himself developed his linguistic model as follows: The contrast signifies two ways of looking at the 'behavior stream': To view the stream etically is to analyze it into elements that fit into something external to the cultural system itself – for instance, if one analyzed a set of practices as 'religious' or 'economic' for the purpose of categorizing them for cross-culturally valid, comparative theorizing. Emic categorization, on the other hand, aims to organize behavioral items so that their interrelation *within* the source-cultural system is revealed. At one point Pike draws an analogy to an automobile: Viewing it emically, we 'might describe the structural functioning of a particular car as a whole, ... showing the parts of the whole car as they function in relation to one another.' Viewing it etically we 'might describe the elements one at a time as they are found in a stock room where ... [they] have been systematically 'filed' according to general criteria' (1954:10). Emic analysis breaks the behavior stream into emic units: concepts, distinctions, etc. as they are viewed by subjects. Instrumental measurements, and more generally discriminations made in terms of imported categories, on the other hand, are etic: While units 'differ etically when instrumental measurements can show them to be so,' they differ emically 'only when they elicit different responses from people acting within the system' (1954: 38). (Note that the methodology and the units of analysis it uses are both termed emic; moreover, since the units produced by emic analysis are intended to be identical to those existing in the culture in the first place, 'emic' generally applies to the phenomena studied as well to certain types of analytic strategy.)

However, Pike intends the etic category to be analogous to the phonetic language and measuring instrument in that it enables approach and approximation of the emic: 'the etic approach is prestructural in that it is used specifically as a first approximation toward reaching an analysis of the emic structure of language or cultural system.' Emic analysis proceeds through stages of etic approximation and refinement (1954: 9).

I have noted that Pike's characterization of emic analysis easily lends itself to the view that its central idea is to 'see things from the subject's point of view.' This is borne out by the following, later characterization that emic analysis 'must analyze that behavior in reference to the manner in which native participants in that behavior react to their own behavior and to the behavior of their colleagues' (1956:661) – a remark that echoes strong sentiments in the foundational work of Boas emphasizing both the value of analysis of cultures 'based on their concepts, not ours' (1943:311) and the deep problems involved in ridding ourselves of unwarranted preconceptions about culture. Kay notes the relevance of this latter concern not only for those interested in cultural uniqueness, but also for those interested in comparison and general theory:

The very provenience of the emic/etic distinction, namely phonology, should make clear that the guiding spirit of an emic approach is to rid oneself of *pre*conceptions about universal structures so that the data may be analyzed objectively to reveal the true universal structures (1970:23).

An important aspect of emic analysis that Pike's car analogy does not capture is that, in effect, the nature of the 'parts' themselves varies according to whether they are seen from within or from without. A social practice or concept might seem similar enough to those sociologically termed religious,' 'scientific,' 'magical' etc. However, characterization of a practice as scientific implies contrasts with religious or magical notions, and the nature of contrast and overlap within the cultural system in question may be quite different. However, it is only the analogy that falls short here, as this feature is clearly central to Pike's concerns to articulate emic analysis.

However, emphasis on 'subject perspective' suggests to many, despite the behaviorist caste of Pike's own thought, a wedding of the emic with the behaviorally irreducible mental. Emic partitioning of the behavior stream reflects the mind's filtering mechanism – one structured in part by culture, in part by nature. But this expectably brings attention to bear on potentially universal psychological features, which in turn may reside in the unconscious – putting the inquirer, potentially, at a considerable remove from the subject's consciously articulated point of view. Now it is rare that inquirers restrict emic analysis entirely to what subjects actually say or believe: even Pike and other behaviorists acknowledge the inaccessibility of certain emic notions to subject awareness, not to mention the fact that subject reports can distort the truth about their beliefs (1954: 9ff.). However, the mentalistic turn entails certain ambiguities that seriously obfuscate emics-etics issues.

EMICS AND MENTALISTIC ANALYSIS

The influential work of Franz Boas (e.g. his 1911) encouraged concern both with understanding the thought of cultural subjects and with the idea that to achieve this, analysis 'must be based on their concepts, not ours' (1943:11); and thus the Boasian tradition is often termed 'emic anthropology.' Yet also in Boas is the recognition that speech behavior, whose study he made so pivotal to anthropological inquiry, is largely governed by unconscious laws (1911:62). The wedding of emic and mentalistic analysis is also evident in the influential work of Goodenough (e.g., 1951, 1956, 1970) and Frake (1962, 1964), which in turn spawned a number of later Boasian approaches variously termed 'ethnoscience,' 'the new ethnography,' 'cognitive anthropology,' etc., which generally ascribe to Goodenough's view that culture is 'the form of things that people have in mind, their models for perceiving, relating, and otherwise interpreting them' (1964:36). Many anthropologists in this lineage are decidedly interested in underlying (often unconscious) cognition and linguistic structure. Given the already heavy linguistic emphasis in the

Boasian tradition, the shift from structuralist to transformational linguistics accounts for a lot of the shift to unconscious cognition; while another Boasian route from linguistics to the unconscious is carved by Lévi-Strauss' exodus from DeSaussure and Jakobson to structuralism (Hymes, 1970: 291ff.). Also, as one would expect, psychology has had considerable impact on cognitive anthropology (Ember, 1977; Kiefer, 1977).

It has been typical of such approaches to develop highly articulated analytic strategies of inquiry, or 'eliciting frames,' that may incorporate depth-psychological testing procedures. However, concerns have been voiced, often by inquirers with general mentalistic leanings, that these approaches have become decidedly 'non-emic': It is argued that ethnoscience has become as 'inquirer-oriented' as structuralism, which despite its Boasian lineage, is rather extreme in its scanning of ethnographic data for evidence of presumed universal mythic and mathematical correspondences (clearly an 'etic' approach in the terms of Pike's automotive analogy). Frake is notable in sounding these sorts of warnings and calling for a return to the general idea of learning how subjects, not anthropologists, organize natural and social reality. He makes the criticism noted earlier that such strategies impose excessive and alien ('platonistic') formal structure (1977; cf. his 1964).

Similarly, while the wedding of emic and mental suggests profitable overlap with psychology, this too will involve 'sorting' cultural phenomena by external constraints. And while it might seem unproblematic to apply psychological universals here, there is a danger that this will after all run against Kay's counsel to avoid finding preconceptions writ large, as psychology itself may reflect ethnocentric bias. (See, e.g., Berry, 1969; Kiefer, 1977; Ember, 1977; Biernoff, 1982.)

EMICS AS ADOPTION OF THE 'TELEOLOGY OF THE OBSERVED'

While 'emic' and 'mental' are equated on many usages, 'emic' more typically designates a methodological commitment to the broad use of translation-based strategies and concepts. This can be viewed as demanding access to subjects' 'inner' cognitive states, but need not be – instead (and, indeed, more in the spirit of Pike) it can be construed in terms of gaining access to significance at the social-behavioral level (see Hymes, 1970:308ff.). The term 'emic' here designates a broad (and still 'Boasian') methodological commitment rather than a specific analytical procedure. It involves not only heavy reliance on translation, but also often inquirers' *use* in ethnographic description of culturally-specific concepts familiar to the people they study – typically in ways that have fundamental structural impact on ethnography or ethnographic methodology. (This involves more than the phonological model suggests: Just as the linguist *uses* phon*etic* notions to describe phonemic contrast, so in semantic analysis, etic notions can be used to determine emic contrast.) For instance, Goodenough has advocated the use of 'emic primitives' in the study of culture. Such terms occupy the fundamental role in analytic frameworks typical of all primitives, i.e., as the undefined bases of definition and analysis; but they also express subject concepts.[3] In his study of the Trukese (1951), he employs as emic primitive the notion of a 'corporation,' which combines relations of property ownership, transfer, and kinship in ways unique, according to the anthropological knowledge of the time, to the Trukese.

Although aligning on one side of a fairly dichotomous methodological contrast, Goodenough emphasizes the aforementioned complementarity of emics and etics in two ways. First, he joins Kay (and Weber) in stressing that *theoretical* understanding of subject perspective requires placing emic understanding in a comparative (viz., for him, 'etic') framework (see, e.g., his 1970). That is, he clearly eschews the 'descriptivist' non-comparativist position that some anthropologists hold – namely, that cultural uniqueness largely precludes comparison and systematic culture-theory construction. In a second, related sense he also speaks of emic *concepts* complementing etic ones, intending the following relationship: Etic concepts are imported in order to enable determination of emic contrast. E.g., in componential analysis (a particular type of contrastive approach), imported technical kinship terms, such as 'collaterality', 'bifurcation', 'descedence', 'genealogical distance', are used to characterize various 'dimensions' of kinship so as to enable the anthropologist to delineate contrastive kinship relations (articulating, for instance, cases in which cultures, unlike ours, find collateral or sexual difference contrastive with respect to relatives to which we would apply the same term 'cousins'). Now this seems an exact model of the phonetics/phonemics relation, namely, etic concepts enabling recognition of emic contrasts. However, Goodenough also speaks of emic notions *becoming* etic notions to be used in the study of other societies in advocating expansion of the inquirer's 'etic kit' for cultural comparison by the inclusion in it of emic notions. A theoretical framework that encompasses more of the known possibilities is more likely to contain conceptual structures manifest in other societies. In fact, Goodenough notes the later use of the corporation concept by another anthropologist studying another culture in the same general region as Truk (1970:70ff., 108ff.), and one can see technical kinship terminology as evolving from emic usage as well. (However, the issues here are somewhat clouded by the terminological point that Goodenough evidently makes cultural uniqueness a *necessary* condition for being emic: thus a notion familiar to members of one culture, but also to members of another is 'etic': so, in light of its cross-cultural instantiation, 'corporation' is now an etic notion. I shall discuss this further later.)

Also, to the degree that Goodenough emphasizes the use of technical idioms in componential analysis, he encourages just the predominance of frame analysis to which Frake (1977) objects. And Frake's methodological response fits well, and probably better, with the idea that emic analysis involves the adoption of subject teleologies. He advocates an emic dramaturgical style of analysis in which key roles in social interaction are discovered by extensive preliminary inquiry in 'query-rich' settings which guide the questioner to ask subsequently the emically appropriate questions (not those comprised in a predetermined eliciting frame). This general idea recurs throughout the literature on more recently emergent approaches such as symbolic anthropology (see, e.g., Douglas, 1973, 1975; Geertz, 1973, 1983) and sociolinguistics (see, e.g., Gumperz and Hymes, 1972), although there is variance even here regarding the importation of predetermined structure. The philosophical background and the methodology of the symbolic school is quite diverse, drawing in some respects on the work of Goffman, in others on that of Winch and Wittgenstein.

In this latter regard, Crick (1976) emphasizes the difficulties involved in the study of 'forms of life' that radically diverge from the inquirer's. The 'map' that an anthropologist draws of a cultural practice must grasp its point, as construed by the subjects, and must be constructed by methods designed to reveal an interconnected conceptual web. There is strong focus here on Pike's idea that emic analysis must show how the conceptual 'pieces' of culture fit together. Thus Crick and other symbolic anthropologists concern themselves with how to penetrate alien culture in a way that permits sensitivity to the possibility of radically different contrastive relations. To do otherwise may be to 'violate some crucial cultural distinctions' and effectively to abandon the whole point of emic analysis altogether: 'There is little point in stressing 'emic' units,' he remarks, 'if domains themselves are unthinkingly thrust on other cultures' (72). Like Winch, Crick draws on Wittgenstein to argue that discovering cultural significance critically hinges on grasping the underlying 'points', and 'language games,' and practice-constitutive rules. With Winch, he argues for a shift of emphasis away from what Crick terms 'scientific mapping principles.'

CULTURAL MATERIALISM

In the spirit of emic analysis, it will help to look briefly at the contrasting methodological emphasis manifest in the 'cultural materialist' perspective of Harris (1986, 1979), from which are voiced broad, and often harsh and controversial criticisms of what he terms 'cultural idealism': a label which embraces all approaches that strive to adopt the teleology of the observed, as well as mentalistically oriented structuralist and ethnoscientific methodologies. Although his criticisms take many forms, their main motivation is a perception that emphasis on subject viewpoints or mental content in any sense threatens theoretical productivity. And here he places a heavy reliance on *pragmatic* bases for theory-construction. Citing the impact of research by philosophers of science such as Kuhn and Laudan, who turn attention to the importance of replacing vaguely articulated concerns to reproduce reality with close understanding of the presuppositions and guiding questions definitive of science and its various paradigm-communities, Harris offers a markedly contrasting view to Frake's with respect to the generation of basic, research-guiding questions. For Harris, such questions reflect the largely antecedent theoretical interests of the general scientific community.

The test of the adequacy of etic accounts is simply their ability to generate scientifically productive theories about the causes of sociocultural differences and similarities. Rather than employ concepts that are necessarily real, meaningful, and appropriate from the native point of view, the observer is free to use alien categories and rules derived from the data language of science (1979:32; cf. 41).

Among the antitheoretical charges he levels are these: that idealists introduce inflexibility by limiting the analyst's fund of basic organizing principles; and that their research tends to concern relatively trivial cultural items, those best amenable to their existing analytic constructs and strategies. (It is interesting that he turns charges of triviality against idealism, many of whose proponents level just this sort of complaint against etically-oriented paradigms.) He also contends that they are inclined to be too ingenuous regarding the veracity of their informants. In contrast to Kay's insistence that emic qualitative analysis is amenable to successful quantitative analysis (Kay cites the aforementioned work by Geoghegan on residence decision rules as a case in point), Harris argues that it is essential to use etic units in such studies, because typical units of analysis such as community organization, family organization, marital residence, etc., are particularly prone to emic/etic confusions (49; cf. Kay 1970: 28f.). (See Marano's discussion of parallel problems in the literature on Windigo phenomena (1982).)

Harris sees himself as operating in a different paradigm, which finds culture within the behavior stream:

> I see Pike's emic/etic distinction as providing the key epistemological opening for a materialist approach to the behavior stream. Goodenough 'sees' emics and etics from an idealist perspective in which the entire field of study – culture – is off limits to the materialist strategies. That is, for Goodenough and other cultural idealists, culture designates an orderly realm of pure idea while the behavior stream is a structureless emanation of that realm (1976:343).

Harris' cultural materialism emphasizes economic and biological factors that cut across cultures. His relatively minimal reliance on translational data is largely restricted to determining gaps between subject belief and factual reality – or 'mystification' in roughly Marx' sense. In marked contrast to idealist concerns to use etic analysis to reveal emic structure, Harris contends that 'etic analysis is not a stepping-stone to the discovery of emic structures, but to the discovery of etic structures' (1979:36). (This bears only superficial similarity to Goodenough's construction of etic kits from emic analysis.)[4]

EMICS AS WHAT IS ACCESSIBLE TO SUBJECT CONSCIOUSNESS

In large part the emic/etics dispute hinges on the amount and kind of emphasis interpretation is given. However, there is another important, and also controversial, feature of Harris' position, namely a different concep-tualization of the key terms at issue:

> whether a construct is emic or etic depends on whether it describes events, entities, or relationships whose physical locus is in the heads of social actors or in the stream of behavior. In turn, the question of whether or not an entity is inside or outside some social actor's head depends on the operations employed to get at it (1976:335).

The interrelation of emic *method* and emic *concept* and *phenomenon* is evident here. If 'eliciting operations,' queries put to the informant regarding the informant's own thoughts and feelings, are necessary to determine the truth of some hypotheses about the significance of some activity, the revealed significance is emic. The informant's reports are viewed as a key source of emic data and the informant is viewed as largely the final arbiter regarding whether the contrastive distinctions the anthropologist makes, and generally the integration of cultural segments, is correct or appropriate. Harris opposes any equation of 'emic' and 'culturally unique'. A notion can be both emic and cross-culturally instantiated if it is 'in the heads' of members of different cultures.

As we saw earlier, Pike views units as differing emically 'only when they elicit different responses from people acting within the system' (1954:38). Yet while this was intended to allow non-verbal behavior to be used as a yardstick of emic significance, Harris is concerned to restrict the 'different responses' to verbal responses to query: 'The way to get inside of people's heads is to talk with them, to ask questions about what they think and feel.' On the other hand: 'Observing what people do during the natural course of behavior stream events leads to etic not emic distinctions' (336). Here Harris' motivation seems more epistemological than materialist-metaphysical: The only way he can make sense of the idea of 'subject' rules 'really' governing subject behavior is by restricting reference to the rules people actually use in the explicit justification or explanation of what they do. Talk of unconscious emic notions, especially as revealed by psychological testing, invites confusion of emic with etic notions. (In this connection see Marano, 1982.)

However, Harris' circumscription of emic significance to what elicitation can reveal is not intended to restrict emics to reporting what subjects say, nor even to make them final arbiters of every inquirer judgment regarding emic significance. The phrase that recurs is that eliciting operations get at what is 'in the head' of the subject, which can allow for things unconsciously so located. For instance, in his 1976 he speaks of two notions of meaning: (1) an etic one, comprising speech acts such as requests for attention, i.e., the 'surface meaning' as comprehended by an observer and (2) an emic content, which concerns 'its total psychological significance for speaker and hearer respectively' (345). Surface meaning can obscure emic meaning, as is evidenced by the fact that frequent but unenforced 'requests' for quiet by a mother of her children might indicate the absence of any intention to get them to perform the requested thing, thus precluding its being a genuine request:

Her failure to take decisive action may very well indicate that there are other semantic components involved. Perhaps she really intends merely to show disapproval. Or perhaps her main intention is to punish herself by making requests with which she knows her son will not comply (346–347).

These emic possibilities, particularly the latter, might not at all be items of conscious awareness. Harris insists, however, that eliciting operations be used in penetrating beneath surface meanings (348).

Harris' 1976 closely ties the emic to the mental nonetheless, and critics have noted this deviation from Pike. However, in later work he makes a four-fold distinction involving the parameters 'etic,' 'emic,' 'behavioral,' and 'mental' (1976:38). As I believe Jorion (1983:55ff.) correctly notes, here Harris equates 'behavioral' with 'descriptive generalizations about behavior'. These are in turn emic when subjects believe them and etic when anthropological observers believe them and subjects do not. On the other hand, 'mental' means 'normative', and the emic norms that subjects acknowledge are differentiated from the etic norms they do not acknowledge but which observers see them as following. More precisely, etic descriptions and norms are those acknowledgeable by subjects via elicitation. Were there reason to think a particular informant could be led to acknowledge some norm, perhaps because other members of the society do, it could be fairly characterized as emic. Tacit emic understanding is a possibility for Harris.[5]

However, talk of the *possibility* of bringing items to reflective consciousness, as well as of retrieving what is 'in the head' may involve us again in problems concerning the status of elements of the deep unconscious. Harris says enough to indicate that he is not to be taken as allowing supplementation or replacement of eliciting operations with psychoanalysis, Rorschach testing, and the like. Greater ambivalence can perhaps be brought about by reflection on generative grammar and semantics which could be seen as appropriately 'based' on eliciting operations. (However, matters are somewhat clouded by the fact that characterization of deep structures with full ontic status involves psycholinguistic and psychological theorizing.) Yet, while not sharply dichotomous, Harris' 'etic' and 'emic' are well discriminable nonetheless.

Harris, expectably, is sharply criticized by cultural idealists not only for his fundamental divergence in perspective, but also for his disregard for the more deeply unconscious dimensions of the concept of emic study (and some also suggest that he limits emics to conscious dimensions more than he actually does). Fisher and Werner (1987), for example, chide Harris for equating subject viewpoints with 'confused' viewpoints (especially given Harris' concern with mystification); but this points more to the basic questions of methodological emphasis discussed in the previous section than to any problems implicit in Harris' usage.

However, Harris' 'etic' *methodological* commitments notwithstanding, his terminology is adaptable to the idea of working from the 'teleology of the observed,' and outside readers of this controversy must be wary of the terminological character of the dispute here. This adaptability is evident, for instance, in the work of Watson (1981), whose general methodological spirit is much closer to Frake. (Indeed, Harris himself notes similarities between his notion of emics and that articulated by the cognitivist Frake (1976:336; cf. Frake 1962: 76).) Commenting critically on his own earlier psychologically oriented studies of Guajiro Indians (studies he terms 'etic' for their external orientation), Watson remarks that the etic categories and strategies he employed inhibited the appreciation of subject perspectives and with that the acquisition of valuable anthropological insights. Particularly, the Guajiro's own strategies of adaptation to city life in Maracaibo were missed and they were made to seem less flexible, less adaptive, and generally less psychologically healthy and well-oriented than depicted on his emic account (453ff.). His emic model, which analyses 'spontaneously recalled personal data' provided in answers to 'open-ended questions' geared to 'the subject's immediate and authentic interests and orientation,' provided quite distinct results (465ff.). These showed, he argues, that his earlier assessment of their seemingly poor understanding of the city reflected instead his own misunderstanding of their different frames of reference, oriented to Guajiro neighborhoods. Also, he came to revise earlier negative evaluations of their tribal background's influence constrained improperly by predefined etic parameters of adaptive vs. maladaptive behavior. The emic model permitted more positive evaluation of the background's importance in promoting individual self-respect.

Watson's notion of emics is equivalent to Harris': Eliciting procedures are used to reveal emic structure, and depth-psychological testing is precluded. Yet in methodological contrast to Harris, he relies on subject schemes for analytical guidance and structure.

EMICS AND INACCESSIBILITY

Harris and others stress the connection between emics and elicitation. However, others stress the element of cultural specificity and uniqueness to the extreme, equating 'emic' with 'untranslated' or even 'untranslatable' (see, e.g., Triandis, 1976: 229f.). The existence of such usages signals a number of the deeper problems, some touching on the very aims of cultural anthropology itself. Evidently, if an emic phenomenon is untranslatable, then it is hard to see how an inquirer will ever express an emic concept in receptor-language or metalanguage terms.

Such usage can be tied to broader commitments to descriptivism, which is typically defended by *a priori* appeal to the *sui generic* character of each culture, but it need not be. Indeed, Goodenough, whose methodology is intended to be as comparative as it is emically sensitive, employs much this notion of emics. Recall that in his discussion of the subsumption of emic notions into an etic kit, the implicit idea was that cultural uniqueness is a necessary condition for being emic, and thus that cross-cultural applicability makes an item etic. Now Harris, who instead emphasizes significance to the subject as the hallmark of the emic, sharply criticizes Goodenough on this point (1976:342). And Goodenough's usage, especially given his comparative aims, does seem to invite undesirable, and unnecessary difficulties. If an emic concept is adequately *translated*, and, indeed, some expression of it correctly employed at fundamental levels of cultural analysis, then cultural insiders and outsiders must surely share the concept. But then translatability makes the concept etic, and it is hard to see how, for example, he can ever *use* an emic Trukese concept like 'corporation.' It seems translatable as such if and only if it isn't emic! Perhaps Goodenough can escape some difficulty by noting that in calling them emic *primitives* he means that they cannot be understood by simple translation into, say, English: Instead, much as with primitives in formal languages, an outsider will have to come to an understanding of quite a bit of Trukese culture before the primitive notion will be properly understood.

However, if universal validity of categories is an objective, it seems better served by Harris' definition. Even if we consider the primitive character of such notions, there seems reason to think that once the outsider has grasped enough of Trukese culture to understand what a corporation is, especially if this means success in stating it in English, that it thus becomes etic. (And if the absence of a simple receptor-language correlate is stressed, we can always make one up – e.g., 'Trukese corporation'.) The point of an emic comparative anthropology is to use categories that ostensibly are *discovered*, not presumed, to be universal – not the result of an antecedent choice of categorization. I think Harris does the program of Kay and others a service in delineating emic from etic in terms of methodological reliance on translation. Otherwise, if we state this contrast in a way which makes multicultural sharing preclude emicity, we encourage yet another way of conflating notions that result from translation of the source-culture with those that do not. Only now they become compressed under the 'etic' label, while mentalism tends to conflate them as 'emic.' The deeper contrasts notwithstanding, Harris' notion seems better to fit the spirit of emic-*oriented* approaches.

EMIC AS NON-TECHNICAL AND NON-QUANTITATIVE

One can easily characterize Harris as fully agreeing to the idea of analyzing culture according to what Crick terms 'scientific mapping principles' – while Crick and Frake seek to derive such principles from the culture under study. Since technical terminology is typically articulated according to such scientific principles, this teleological contrast blends easily with one along a technical vs. non-technical dimension (Jorion, 1983:63). The original phonological contrast embodies the idea that etic description, much like phonetic description, is couched in highly technical language, understandable perhaps largely only to the analyst. Technical terminology (and measurement apparata) *enable* the inquirer's eventual discernment of phonemic differences; and similarly, as we saw, for the componential analysis of kinship.

The etic/emic contrast here is fairly distinct from those articulated by either Goodenough or Harris, neither of whom seems to preclude the possibility of emic notions being technical. Nor does it run exactly against Frake's grain, though the 'quantity/quality' worry of distorting social reality with scientific abstraction motivates some to reject technical analysis on these grounds. Also, there is the more essentially 'emic' concern that the technical idiom may spring from fundamental conceptual contrasts quite distinct from those underlying subject discourse, a distortion symbolist critics fear is easily missed because the technical apparatus can carry with it glib acceptance of the idea that its use guarantees 'scientific objectivity.' Moreover, some of the inner tensions of emic anthropology emerge in this connection when the untranslatability of emic concepts is reconciled with their adoption by inquirers by the homophonic subsumption of source-language expressions – witness 'totem', 'tabu', 'windigo', etc.: Thus subsumed, these terms can take on distinct technical meaning, and utterly fail to 'express their original meaning.' (See, again, Marano 1982 on the etic psychoanalytic subsumption of the windigo concept.)

The equation of 'emic' with 'non-technical' suggests also an equivalence to 'non-quantitative'. Indeed, idealists and materialists alike have characterized statistical comparison as ineluctably etic. The idea is partly suggested by Pike's idea that emics works from within the culture being studied, while, as Jorion puts it, 'quantitative or statistical pertains to the bird's eye view,' and 'is derivative of our interest in collective behaviour ...' (1983:49). Thus, as with 'technical' strategies generally, quantitative analysis can be seen as the articulation of scientific, not culturally valid, mapping principles. However, it is not clear that quantitative analysis *per se* reflects choice of deep, culturally alien principles: surely counting is cross-culturally instantiated. Once the system has been revealed, what prevents using in quantitative analysis units whose linguistic vehicles translate source-language expressions? So using them would not seem, *ipso facto*, to 'transform their meaning' in any drastic way (though the more articulated conceptual background might well manifest divergence in underlying 'mapping principle' – but it would seem that translation could provide warranted reason to put such concerns to rest.

EMICS AS INTERCULTURAL 'COMMON GROUND'

Jorion (1983:57) identifies another usage of 'emic' – the contrary of the idea that emic phenomena are untranslatable or resist comparative study – which he terms the 'common ground between Us and Them.' Jorion sees Goodenough as a useful point of departure in this connection, not only because of Goodenough's general emic-comparative aims, but also because of his views about the basic character of culture. In line with his view that culture is comprised of the rules of appropriateness, or for 'getting along' in a society, Goodenough defines the 'problem of ethnography' as 'how to describe a culture of another people for an audience that is unfamiliar with it' and to enable them to understand it as well as its participants do, that is 'in terms that permit discussing it knowledgeably with them' (1970:105). Jorion sees in this the signal of concern that emic analysis reveal 'the intersection (in terms of set theory) of Our understandings and Theirs.' (Interestingly, as Jorion notes (62), the earlier-noted process of absorption of foreign terms, such as 'totem', into 'etic' technical parlance can come full circle in their adoption in 'emic' non-technical western discourse – becoming 'emic' also in the present sense.)

Yet while this usage seems so at odds with the idea of emics as culturally specific, it emerges from the tensions exposed by, to adapt Kuhn's term, the 'essential tension' in emic comparativism. How else are we to guarantee that cross-cultural comparison reveals rather than imposes similarities, except by believing that certain features are culturally shared? Indeed, if one adheres to Goodenough's program of expanding the 'etic kit' – but joins Harris in terming it an 'emic' one, if it succeeds in finding 'the same things in the heads' of members of different cultures – then anthropology not only springs from this 'common ground,' but aims to expand it (but in a less 'imperialistic' or 'hegemonistic' way, some would say, than shameless impositionists like Harris).

However, the existence of such conceptual intersection is questionable, as is evidenced not only by the challenges to emic anthropology but by its internal tensions. Kay and Goodenough insist that the overlap must be discovered, and not be the product of imposition. Yet commitment to the idea that at various deeper cognitive levels there already *is* overlap is what leads many to see emic analysis as going deeper than what is said or consciously believed. In much structuralism and cognitive anthropology there is the implication that by 'looking within the western mind' psychologists have already discovered universals. Perhaps the boldest statement of this idea is in Lévi-Strauss' remark that (in the analysis of American Indian myths) 'it is in the last resort immaterial whether ... the thought processes of the South American Indians take shape through the medium of my thought, or whether mine take place through the medium of theirs' (1969:13). Indeed, Lévi-Strauss and others examine cultural data in search of deeper cognitive overlap or psychic unity (based ultimately in brain structure).

Now the obvious response of those who are more concerned to discover universals at the socio-cultural level, or to look more carefully at that level for real evidence of deeper universal sharing (perhaps because they fear that psychology itself has ignored cultural divergence) is that psychic unity must not be *assumed*, but rather genuinely discovered. But here important epistemological problems intrude. Particularly, there is the question of whether and to what degree cognitive overlap is discovered rather than presupposed if the task of 'making sense' of other cultures is to get off the ground at all. Hollis, for instance, argues that certain minimal criteria of rationality must be imposed on subjects as a matter of synthetic *a priori* necessity (1970a, b). For some, a somewhat stronger objective unity of belief or cognition can be derived from the possibility of communication or interpretation (Davidson, 1983, 1984a,b; Lear, 1982).

Jorion notes a rather strong commitment to psychic unity in the non-anthropological setting of Strawson's views concerning philosophical explication – which are strongly reminiscent of Lévi-Strauss's views regarding the timeless, ahistorical character of mythic substructure. According to Strawson, there exists 'a massive central core of human thinking which has no history,' comprised of 'categories and concepts which, in their most fundamental character change not at all' (1959:10; cf. Jorion, 1983:59). For

Strawson these pervade all discourse, from the technical to the ordinary, and their revelation, through philosophical conceptual analysis, defines his program of 'descriptive metaphysics' – the Kantian program given a linguistic turn.

However, the breadth and ahistorical character of Strawson's and Lévi-Strauss' conceptual cores creates the fear that anthropology may become all too easy. If its presumption is without warrant yet also that which guides what inquirers look *for* in culture, it may well be only pervasive dogma, granting anthropologists, to cite a relevant quip of Russell's, all the virtues of theft over honest toil.

THE INDETERMINACY OF TRANSLATION

The question of presumed common ground may well be life-threatening for emic anthropology – if imposing the familiar precludes the recovery or use of cultural kinds by the inquirer. For some degree of imposition seems intrinsic to the whole ethnographic idea of making sense of a culture to an audience of outsiders. And if not fatal, the problem at the very least places clear burdens on cultural idealists to clarify their aims and claims. While its proponents may disagree with Harris' views, they should take seriously his call in the early pages of *Cultural Materialism* 'to replace the inchoate and unconscious paradigms under whose auspices most anthropologists conduct their research with explicit descriptions of basic objectives, rules, and assumptions' (1979:26).

The key problem is put in most perspicuous form in W. V. Quine's 'indeterminacy of translation' thesis which sees the insufficiency of behavioral evidence to fully warrant translation as entailing that there is 'no fact of the matter' to translational choice. I shall consider the import of this thesis specifically in relation to the 'emic' approach advocated by Watson (which applies Harris' definition of 'emic' analysis while diverging from him on methodological emphasis), since it defines a distinct and potentially fruitful methodological commitment. That is, emic analysis is analysis that restricts itself to use of conceptions that subjects can use, without extensive technical training, in the explanation of their own behavior. It is distinguished (though in no sharply dichotomous fashion) from etic modes of analysis that use predetermined strategies and terminologies that are not based on the standard translation of the discourse of the society under study.

Quine's immediate target is a wide range of epistemological and semantic assumptions prevalent in traditional and more recent positivist and ordinary-

language philosophies, among them, the descriptive metaphysics of Strawson. Particularly, Quine challenges the idea that the conceptual substratum of
language can be revealed by linguistic analysis (whether done by linguists or
philosophers). In his view pragmatic criteria of adequacy guide these efforts,
not any semantic fact. Ordinary usage generates very little by way of
determinate, antecedently existing synonymy relations for such analysis to
reveal (1960:26f.; 1979:166f.).

Most important here is a line of argument Quine uses to support this
critique which involves an anthropological thought experiment designed to
show the limits of what can be ascertained in translation. Roughly, the idea is
that if underlying 'natural synonymy' relations really are the factual standards of interpretive correctness (whether it be translation or explication),
then a linguist translating the discourse of a hitherto untouched culture,
bearing no similarities to existing cultures (an idealization), with all possible
behavioral data at his disposal (another idealization), should be able to get
determinate translational results. However, Quine argues, he cannot. In terms
of Pike's insistence that 'units are different emically only when they elicit
different responses from people acting within the system,' Quine's charge is
that even in light of all imaginable evidence divergent translations will differ
sharply from each other but will not differ emically (1960: Chap. 2; cf.
1951b). Indeed, though radical translation is an idealization – at best
approximated in anthropological field translation – aimed at showing the
inability of conceptual analysis (a kind of intralingual interpretation) to find a
timeless core of thought, it also aims to show the core's inaccessibility
(whether by direct or transcendental means) to interlinguistic translation. And
taking this all in the intended verificationist spirit, we are to conclude that
there is no such core.

Consider this widely-discussed example: We would suppose the terms
'gavagai' and 'rabbit' express the same concept, or refer to the same object,
were we to have isolated and matched the stimulus conditions under which
speakers affirm and deny the *sentences* 'Gavagai' and 'Lo, a rabbit'; or, as
Quine puts it, were we to have discovered that the sentences have the same
'stimulus meaning.' However, the same evidence – no matter how much of it
we garner – also supports the equation of the term 'gavagai' with 'rabbit
stage', 'undetached rabbit part', and indefinitely many other terms, all of
which refer to different kinds of thing. All that is needed to enable behavioral
evidence to support these odder alternatives is compensatory adjustments in
our analysis of the source-language's individuative apparatus – the articles,
identity predicates, pluralization devices, etc. The imposition of such

'parochial' (and 'etic') constructs is necessary here, they are not 'given' in the behavior stream; but there is reason to doubt that any 'emic approxima- tion' can ever be achieved through their use (51ff.). At this basic level it makes no sense to speak of approximating something.

The 'gavagai' problem specifically challenges the translation of terms, and Quine sees indeterminacy as pervading the translation of most sentences as well. The latter, broader thesis follows parallel reasoning: All behavioral evidence leaves us with a multiplicity of divergent right answers. Additional criteria are needed to narrow the field, but these then take us beyond what the evidence avails us by way of clues to what *subjects* mean or refer to. In effect they amount only to *imposition* of concept and belief to make for easier and more fluid grasp of the source-language by *users* of the translation manual. Appealing to a holist thesis first articulated by Pierre Duhem, Quine argues that we have great freedom of choice in manipulating scientific theories in the light of experimental anomaly: to the degree that any hypothesis can be saved if compensatory adjustments are made elsewhere. This means that most statements in a theory cannot be tied to some specific set of testing observ- able consequences, but that they face the 'tribunal of experience' as parts of larger theoretical systems of belief (1951a). The linguist, in turn, has similar freedom in employing receptor sentences as translations of source sentences: The receptor component of a translation manual is a systematic account of physical and social reality: a theory of nature (in a broad sense), ostensibly the subjects'. Now consider a sentence S in the source language which cannot be translated simply on the basis of stimulus conditions of assent and dissent, and assume it is translated as R in a successful manual of translation. Owing to holism, S can be translated as the distinct R´ in another manual, providing compensatory adjustments are made elsewhere: that is, R´ will be embedded in a different system of interrelated sentences that serves as the receptor component of the second manual and comprises a distinct theory of nature. Both 'S means R' and 'S means R´', and their containing manuals of translation will square with behavioral evidence. But they can give sharply divergent interpretations, 'perhaps patently contrary in truth value' (1960:72f.): The manuals are in no sense equivalent, yet no behavioral evidence can serve to distinguish one as the uniquely best one (1969:80ff.)[6]

Of course holism pervades natural science as well, but occasions no indeterminacy of scientific hypotheses or theories. However, Quine maintains that the methodology of translation does not adequately answer the question 'what is our behavioral evidence for ascribing particular meanings to source- language expressions?' The difference lies in the following considerations:

(1) We have no higher standard of truth than what correct physical theory can tell us. Thus the possibility of alternative and divergent theories squaring with the evidence cannot give reason to challenge theoretical truth at that level, since it would involve use of an illicitly 'transcendent' notion of truth. (See Gibson 1987, Quine 1987) (2) Even having settled on a physical theory, even in light of all imaginable evidence, we must rely on supplemental considerations in selecting translation manuals. (3) These considerations involve imposing familiar schemes of reference and assuming that subjects hold beliefs that are obviously true to us. Though not translatable on the basis of stimulus meaning: that is, by operating under a 'principle of charity.' But in so doing we make subjects out to share our concepts and beliefs as a precondition, not a result, of translation. What sense, then, Quine asks, in thinking that there are antecedent semantic facts, synonymy relations, embedded meanings, that serve as the standards of correct translation? What emic structure is being approximated here?

Now just how these three points apply is not clear, much less whether the thesis is convincing in any case. Indeed, while Quine gives much attention to interpreter imposition, and while the context of emic anthropology would suggest that this would be a telling concern with respect to interpretive objectivity, Quine at times sounds as if he would be content to generate the thesis from the first two considerations alone. Roughly, Quine's idea is that there are no differences in fact if there are no determinate differences in physical state, and translation suffers indeterminacy because distinctly divergent translation manuals can be physically equivalent (1970; 1981). However, it is not clear that Quine's 'physical hegemony' is warranted (see, e.g., Goodman, 1968: 262ff.; Rorty, 1982: 201). My own view is that interpreter imposition is more integral to the indeterminacy thesis.

Davidson, a student of Quine, shares my concern that imposition be given more play, and as his work has had some considerable impact on social science, and as it largely expands on the Hollis position sketched earlier, I shall give it some brief mention. Davidson believes that there is an indeterminacy of translation, but that it (and the attendant indeterminacy of emic qualitative analysis) is a harmless feature of all qualitative analysis: There is similarly no fact of the matter to our choice of temperature scales, and we can choose to measure in Fahrenheit or Centigrade units without fear of compromising factual objectivity. Basically, while he joins Quine in insisting that charity is essential, that translation is not possible unless extensively shared belief is *presumed* on the part of the parties to translation, he believes this is to be fit into an account of interpretive objectivity, not to be seen as grounds for interpretive skepticism. The crucial point for Davidson is that we *must* be charitable if we are to make sense of any of the claims at issue. Belief in intercultural conceptual variation must be warranted by successful translation: the possibility of divergence undetectable in translation lacks empirical basis. Contrary to Kuhn (and cf. Sapir, 1964:128), total or near-total incommensurability is not discoverable, if it is tantamount to untranslatability (Davidson, 1983, 1984a, b).

It is worth adding that Davidson tries to go beyond this Hollis-style epistemological line with what Larson (1987) terms an 'objectivity thesis': He tries to establish objective determinations of correct or incorrect interpretation, and to show that such determinations are possible only against a setting of largely *true* belief (not simply belief held true by inquirer and subject). Briefly, since Davidson can make no sense of interpretive error independent of what interpreters could discover, he appeals to an 'Omniscient Interpreter' in explaining the actual correctness or error of fallible human interpreters: 'The omniscient interpreter, using the same method as the fallible interpreter, finds the fallible speaker largely consistent and correct. By his own standards, of course, but since these are objectively correct, the fallible speaker is seen to be largely correct and consistent by objective standards' (1983: 435).

INTERPRETATION AND OBJECTIVITY

Does Davidson's solution adequately address the question of interpretive objectivity? For one, the focus on translation leaves the concern that belief in intercultural conceptual divergence beyond what is evident in translation may be based on a *bilingual*'s ability to grasp two divergent systems but be unable to translate considerable stretches of one into the other. Indeed, total failure of translation might be conceivable (Quine considers viable just such a possibility in his 1987:157), and this casts some doubt on both Davidson's epistemological and objectivist theses. Even the omniscient interpreter might fail to effect translation in such a case. For this and other reasons there remains some question as to why Davidson's *objectivist* conclusion is merited by considerations of *imposition*, as opposed to something somewhat more skeptical.

Quine seems to have more skeptical import in mind, but is is not clear what it is. Why is there 'no fact of the matter' to translation, and what does this entail, exactly? Are emic methodologies that *rest* ethnographical justification on translation at hazard because the latter is indeterminate? Is denying factuality to its premises fatal to its fundamental objectives? I think the answer to this last question is no, even though translation is indeterminate in a special way that results from interpreter imposition. Moreover, interpretive objectivity can be explained without trying to assume any metaphysical, supratheoretical perspective. Now in part what I have in mind here involves general questions of truth and objectivity that cannot be resolved here. However, I will indicate what seem to me promising lines of argument.

Subsequently, I shall address the translation-specific issues to which Quinean indeterminacy turns our attention.

Turning to the general issue, we have seen Quine appeal, in roughly Peircian fashion, to physical theory as an ultimate ontological standard: there are no differences in fact without differences in physical state, something which seems to entail difficulties for translation. Davidson, on the other hand, tries to guarantee translation's objectivity by calling attention to what an omniscient interpreter could know about true beliefs actually shared by the parties to translation (and necessarily imposed by one on another). Now such efforts at defining and assuring objectivity work, effectively, from within our existing scientific standards, and traditionally invite Cartesian worries concerning the possibility of global error, or generally concerning failure to *justify* thinking our best efforts to portray reality actually do so. In response, idealists and pragmatists, who have typically articulated such internalist positions, have tried to attack the sensibility of such externalist justifications, along with that of the skeptical worries that motivate them. Particularly enlightening in this latter regard is the 'irrealist' thesis advocated in Nelson Goodman's recent work (1978: Chaps. 1, 6, 7; 1984: Chaps. 1–2). Goodman employs strategies reminiscent of idealism (particularly the broad semiotics of Cassirer, whom he favorably cites (1978: 1ff.)), but his arguments are particularly valuable in that he works against much the same background as Quine and Davidson: a pragmatist critique of the notion that philosophy aims at the articulation of conceptual schemes, and of the host of epistemological and semantic concepts and theses that attend this attitude. Indeed, all three try to articulate the basic thesis that a sharp distinction between a conceptual scheme and the content it organizes cannot be maintained, and that this has significant import for our views of objectivity, justifcation, and the function of philosophy. I think Goodman's advantage lies in giving more attention to challenging our externalist-metaphysical intuitions on these matters. (Indeed, it goes as far as challenging, perhaps, the very motivation for Quine's physicalistic ontology and Davidson's omniscient interpreter.)

Pointing to much the same embarrassment of riches as Quine – divergent yet equally adequate alternative 'versions' of the world, Goodman argues against the idea that belief in a transcendental basis for adoption of theories or their categorial ('natural-kind') schemes can be given philosophical support. ('Version' is chosen to stress the equal footing of science and art.) The key insight to be drawn from the collapse of the scheme-content distinction, he argues, is that convention is largely constitutive of adequacy: Natural kinds are 'conventional kinds' – adequate if theoretical purposes are served (while representationalism in art is similarly guided by convention and not 'natural resemblances'). In most general terms, untenable presumptions of comparability of versions to an uncognized or neutrally characterizable 'world itself' underlie skepticism, foundationalism, and most metaphysics (1973: Chap. 3; 1977: esp. Chaps. 1, 6, 7; 1984, esp. Chap. 2.). Questions of objectivity and adequacy are questions that must involve the comparison of versions with other versions. Against both foundationalist and skeptic he holds that the possibility of multiple but incompatible accounts points to *pluralism* of 'objective' accounts, not the absence of any. (Cf. Geertz's favorable citations of Goodman on this score (1983: 118f., 151ff., 180f.).)

Now this is well-traveled and treacherous territory here, and I can hardly offer sufficient grounds to persuade readers to see pluralism rather than skepticism as following from these concerns.[7] However, these are considerations worth careful attention by emic anthropologists interested in freeing their methodologies from commitments – that threaten to be most untenable – to the reproduction or mirror-reflection of social reality (see Rudner, 1973). Emic anthropologists must be wary that their desire to recover and employ as units of analysis the 'cultural kinds' of their subjects doesn't founder on a problematic assumption that a supraparadigmatic vantage point may be occupied – an assumption possibly founded on a likewise problematic view that the validity of natural-kind schemata in science rests on warrantable comparisons with unsymbolized or neutrally-symbolized reality.

RECONCILING IMPOSITION WITH EMIC OBJECTIVES

Let me turn now to matters specifically pertinent to translation. While I believe Goodman provides valuable resources for addressing certain general aspects of the question of 'fit' at issue here, there remain some specific concerns raised by indeterminacy. Might we not say, from *within* our theoretical framework, that Quine raises doubts about the coherence of emic anthropology. The question, again, is this: If only imposition overcomes our embarrassment of interpretive riches, how can we trust emic methodologies that *rest* their factual case on indeterminate premises, on assumed cultural overlap that can in no way be justified as even 'approximately emic'?

In response, I believe we should see the imposition problem as pointing to the fact that translation is not description of meaning, but a codificational enterprise aimed at *enabling* and *facilitating* cultural interaction and anthropological theory-building (see my 1982, 1986, 1988a). Learning the meanings of expressions is no more than formulating enabling conventions. The imposition problem bears out their non-hypothetical character, and the temptation (which should be resisted) to view them as failed hypotheses stems only from certain similarities in logical character between prescriptive and descriptive systems.

The source-language community confronting the interpreter will show certain rule-governed regularities in linguistic and related non-linguistic behavior. These may themselves have been subject to codifications such as the formulation of a written language, with written dictionaries and grammars, or the rules in question might simply be followed tacitly. (We can, with Wittgenstein characterize rule-governed behavior without reference to explicit or statable rules. It can be regarded entirely in terms of regularities in behavior in which errors can be committed by individuals who can then in turn be corrected.) The translator's efforts are aimed at bringing other 'players into the game,' by mapping, on the basis of stimulus meaning or grasp of a sort of equivalence of function, receptor-language expressions onto source-language ones. The ultimate measure of these efforts is effective communication and general interaction by outsiders with cultural insiders. 'Parochial' imposition is to be expected in translational conventions because they are primarily a guide to the outsider, a set of rules of translating behavior that facilitate linguistic interaction: Their peculiar feature is that they guide only the manual user; the insider with whom the user interacts is following different rules, statable, if at all, in the source language.

Following the albeit vague usage that has arisen in the relevant literature, I use 'rule' broadly here to cover conventions of various sorts. Conventions are regularities in behavior meeting the following conditions: Nearly everyone in the community conforms to the regularity in some recurrent type of situation and expects others to do so. However, non-conformity must be a possibility (in the sense just described) and conformity by subjects must be conditional on the warranted expectation that others will similarly conform. Mutual expectation of conformity stems from belief that others will recognize that it is in their self-interest to conform, if most others do; and in certain instances such expectation may be reinforced by additional expectations of some form of sanction for non-conformity (and some may prefer to reserve the label 'rule' for conventions meeting this latter condition). For instance, in certain societies there is a convention of walking on the right side of a path, which most conform to because they expect that others will do so because they in turn expect most others will do so. Deviations are possible, as are alternative conventions – walking on the left side – if mutual expectations support them. It is just expectations formed in the belief that all view themselves as better off (here because collisions or forced sudden deviations in direction are minimized by conformity) that guide individuals. Meanwhile, in the case of *driving* on one side or the other, greater disutility of non-conformity makes it reasonable to impose sanctions against non-conformity. Linguistic 'rule-governed' behavior is similarly bolstered at least by expectation of mutual benefit and often by some (typically mild) form of corrective action or sanction.

These considerations of function are supplemented by reflection on the logic of translational compliance – which seems to me to better approximate rule-compliance than descriptive law-compliance (although, again, both sorts of compliance bear many similarities to each other). Roughly, what distinguishes rule-governed from natural-*law*-governed behavior is the fact that behavior can conform to a rule without thereby abridging it, but not so with a law. A counterinstance can be ignored, relegated to probable experimental error, or taken seriously – but it constitutes an anomaly if allowed to stand. By contrast, violability without abridgment is a necessary condition for something's being a rule – yet by the same token there are also fairly clear conditions under which counterinstances do abridge rules: when violation is sufficiently widespread, or when people particularly competent in a practice find following the rules does not best facilitate the practice in question.

Generally, prescriptive systems succeed when they properly codify the practices from which they are derived. A prescriptive system of rules of hypothesis acceptance arises from articulation of statements of conventions that competent practitioners follow, and the point of codification by reflective formulation of governing principle is to refine and improve the practice. Much the same goes for codifications of legal systems. Refinements in law are evaluated by the competent practitioner's intuitions regarding the nature or intent of the earlier unformulated or less adequately formulated principle. Similarly for linguistic practices: grammars and dictionaries are derived from pre-existing practice and sharpen and facilitate the practice. Deviations will occur after codification, perhaps ones that would not have been so clearly perceivable as such without benefit of the codification. But such deviations need not abridge the rules in question. However, again, sufficiently serious ones will. If competent practitioners find some philosophical explication of a 'covering-law' model of scientific explanation such as to exclude inferential practices they find acceptable (or include ones they do not), the principles change, not the practice. However, it is largely the principles that are used to decide whether behavior, by the competent or the incompetent, conforms. Successful codification consists in principles being in what Rawls terms 'stable reflective equilibrium' with the practice and intuitive judgment of competent practitioners (1973: 21; cf. Goodman, 1973: 62ff.). (The contrast with *probabilistic* law-systems is not easy to state, but contravening ensembles of events can be well defined statistically, while no such precise definition can be given for rule-contravention: Even though suitably broad or serious conflicts of rule and practice abridge the rule, and even though such conditions are definable, such definition is not a quantifiable matter.)

And so it is with translation. Although here matters are complicated by the fact that the practice – intercultural discourse – is governed by distinct sets of rules, and translation constitutes the rules for outsiders only. Yet the rules are nonetheless sensitive to the practice of cultural insiders as well as outsiders. If there are occasional deviations in usage by insiders speaking their own language, no change in translation is entailed. But if the deviations are serious enough, whether because the original translational task was not done correctly or because usage patterns in the source-culture have changed, the translation rules will have to be modified. Manual users will find themselves not acting in optimal conventional ways, and will change translation rules to overcome this. Translational modification my also reflect changes in receptor-language usage. Thus translational success is really a matter of translation rules being in stable reflective equilibrium with linguistic practices of people who can and also cannot understand them.

Some cultural outsiders are anthropologists, and it is worth also seeing how all this is related to their community's behavior – i.e., how translation facilitates anthropological theory-building. This community, for example, enforces conformity to a wide range of rules of linguistic analysis (including those that impose grammatical categories) in any translation. The thrust of these rules is to maintain behavior-patterns in inquirers that ideally promote the search for knowledge, through the avoidance of unnecessary conceptual diversity in the various apparata of linguistic analysis. Indeed, the interpreter's work facilitates his ability to get along in this community's efforts at theory-building as much as it facilitates his dealings with the source-society. And change in anthropological method and theory is another factor that might destabilize the reflective equilibrium of translation manuals.

It will be helpful in this connection to note the similarities between translation and definition. Systematic definition, a mapping of words onto words, enables all scientific inquiry. It is largely conventional, meaning that different systems of definitions can suffice for those that are embedded in any given successful theory – indeed, recall that Quine's concerns about translation are intended to reflect on even such 'translational mappings' in one language or formal system. In fact, alternatives can coexist, as with the variously defined units of temperature. Further, as we saw Davidson note, the difference, and the non-factuality of the difference, between two characterizations is innocuous. In this respect, finally, I agree with Davidson. But now we can better accommodate our intuitions that interpretation lacks a fact of the matter without fear that the 'radical-translation' thought experiment also threatens field translations that approximate Quine's radical idealization.

I also noted some parallels between translation and codification of rules of theory- and hypothesis-acceptance that govern scientific behavior. These too are helpful in reconciling translation's 'non-factual' status with the potential objectivity of emic anthropology. These principles also, in their way, enable science to perform its predictive and explanatory tasks, while they are not themselves descriptive in nature.

EMIC REALITY

There are similarities and differences between the relationships, respectively, of laws to governed instances and rules to governed instances. The similarities account for the fact that indeterminate translational correlations seem to be disconfirmable descriptive statements. Similarly, success for prescriptive systems is much like success for descriptive theories. 'Disconfirmation' – whether due to lack of attention to proper usage, or due to a general shift in usage patterns subsequent to translation – is a destabilization of that equilibrium that calls for revision of translation rules that outsiders follow in order to prevent resulting failure of optimum coordination of activities (communicative or otherwise). Behavioral deviations contravene both; but the differences in the character of the relations involved in rule- vs. law-contravention are significant.

The relation of translation to definition does not itself encourage regarding translations as rules. (Definitions can be viewed as descriptive in form.) The relation to rules of acceptance does more in this vein. However, regardless of the canonical form one gives translations, the important thing is to see that we can reconcile the indeterminacy of translational correlations, their failure to be genuine hypotheses, with their potential scientific character. Translations are 'basic' to interpretive description and explanation, but they are not basic *premises* in such accounts. We don't have to worry that their lack of truth value compromises subsequent conclusions about culture.

The general thrust of my position is to give just the turn to basic methodological questions that Harris emphasizes, to pragmatic matters of theory construction, to how criteria of adequacy meet clearly identifiable purposes, etc. Higher levels of generality can be achieved by considering interrelations among theories, paradigms, and disciplines – but at no point do I see a need to prove the worth of the criteria and methodologies of anthropology (and social inquiry generally) by recourse to metaphysical theses about culture's essential nature, the psychic unity of mankind, and the like. There is every reason to be a pluralist about ethnography, to embrace the etic and emic accounts, and the many divergent accounts within each camp. Or at the very least, if our concerns stem from a belief that there is, after all, '*a* way the world is,' that this belief is in serious need of justification in light of Goodman's arguments in support of the idea that there are as many worlds as there are adequate versions (cf. his 1972).

This reconstrual has methodological import only in endeavoring to dispel pseudo-problems of objectivity, and perhaps to take the charge out of a number of methodological controversies. It does not purport to change methodology, save for getting it out from under the influence of misguided metaphysics. In a Wittgensteinian spirit, this reconstrual 'leaves everything as it is.' Moreover it leaves emic anthropology a feasible enterprise. Finally, it serves to respond in the following way to Brown's concerns, noted at the outset, to make sense of the idea that such things as kin terminologies 'fit' the reality they are used to describe': First, we must realize the problems of talking about fit whether in general or with respect to translation-based anthropology. With Davidson we ask what 'fit' means in the light of interpreter imposition – or, better, translation's coordinative character: Challenge to interpretive fit cannot be based on a desire to reproduce social reality, and cannot make interpretation impossible. If we look at how facts actually figure in interpretation, we see that translation enables and warrants determinations of fit. Also we must keep in mind Goodman's key challenge to the idea of determining fit of symbolized to unsymbolized reality. This undermines Brown's analogy: According to the irrealist thesis, one never gets anything out of comparisons with 'mute' reality; it does not divide itself into appropriate units. Instead what is operating in the physical example Brown offers is a *prior* systematization. We know not to weigh heavy things in ounces or light things in tons because we have warranted beliefs, statable in scientific or common-sense background systems, to the effect that these are inefficient procedures. Discursive knowledge stated in existing symbol systems guides us here. And thus it is with interpretive systems: the fact that success has been achieved using certain 'community' rules in describing culture is all the success there can be. If we rely on translation at all, we can no more expect a 'pure' account of another system unsullied by the conceptual apparatus of the inquirer than we can expect description to give a pure reproduction of anything it describes. Distinctions between social fact and fiction are made within our systems using the criteria we find acceptable. There is no way to legitimate criteria in any transcendant way – but neither is there a clear transcendant basis for challenging them.

This account also leaves emic anthropology with purpose. It better enables communication with and participation in societies with divergent cultural backgrounds, and avails us of different understandings of societies that might reveal strengths that etic analyses might miss, as in the case of Watson's study of the Guajiro (which is not to devalue, however, the possible insight that psychological-etic analysis might give regarding potential disfunction in their adaptive strategies). Moreover, we can now see more clearly how such expansion of the inquirer's conceptual scheme results from a factual understanding of other societies. Insofar as the codificational nature of translation in no way compromises the objective-descriptive capacity of emic ethnography, we can empirically warrant our belief that we indeed have (hopefully, various) versions of the ways their worlds are.

NOTES

[1] I discuss various aspects of this problem in more detail in my 1982, 1986, 1988a, 1988b. Also, I do not intend to give an exhaustive account of all usages of 'etic' and 'emic'. The reader is referred to the cited papers by Harris (1976) and Jorion (1983) for somewhat different relevant schemata.
[2] Generally, the context of Evans-Pritchard's and Winch's critique of social anthropology is a fruitful source of insight into emics/etics-quality/quantity overlap (Crick, 1976: esp. Chaps. 5–6).
[3] Whether the notions are or need be primitive in the Trukese system itself is not clear.
[4] Jorion articulates a separate category of 'emics' as involving concern with native decision theories. However, I think this is sufficiently similar to the categories I employ here that it does not deserve separate mention. Also, it is worth noting that studies of decision-making are subject to much the same 'emics/etics' controversy regarding the degree to which translation figures in the attribution of decision

strategies to subjects. (See, e.g., Gladwin, 1975; Quinn, 1978).

[5] Harris approximates the notion of 'internal understanding' in Collingwood (1946), Winch (1958), Hart (1961). Winch's Wittgensteinian emphasis, however, may lead him to seek an internal understanding that requires mastery of irretrievably tacit rules.

[6] Cf. the somewhat different thesis offered in his 1970, and cf. his comments on this in his 1987:460.

[7] It is interesting (though perhaps frustrating for the Goodmanian pluralist) to see Kripke's (1982:13ff.) application of Goodman to skeptical ends.

BIBLIOGRAPHY

Berry, J. W., 1969, 'On Cross-Cultural Comparability,' *International Journal of Psychology* **4**, pp. 119–128.

Biernoff, D., 1982, 'Psychiatric and Anthropological Interpretations of 'Aberrant' Behaviour in an Aboriginal Community,' in Reid, J. (ed.), *Body, Land, and Spirit* (St. Lucia: University of Queensland), pp. 139–153.

Boas, Franz, 1911, *Handbook of American Indian Languages* (BAE-B40, part I; Washington, D.C.: Smithsonian Institute).

Boas, Franz, 1943, 'Recent Anthropology,' *Science* **98**, pp. 311–314; 334–337.

Brown, C. H., 1974, 'Psychological, Semantic, and Structural Aspects of American English Kinship Terms,' *American Ethnologist*, **1**, pp. 415–436.

Collingwood, R., 1946, *The Idea of History* (Oxford: Oxford).

Crick, M., 1976, *Explorations in Language and Meaning* (New York: John Wiley and Sons).

Davidson, D., 1983, 'A Coherence Theory of Truth and Knowledge,' in Henrich, D. (ed.), *Kant oder Hegel?* (Stuttgart: Klett-Cotta), pp. 423–438.

Davidson, D., 1984a, 'On the Very Idea of a Conceptual Scheme,' in *Inquiries into Truth and Interpretation* (Oxford: Oxford), pp. 183–198.

Davidson, D., 1984b, 'On the Method of Truth in Metaphysics,' *loc. cit.*, pp. 199–214.

Douglas, M., 1973, *Natural Symbols: Explorations in Cosmology*, 2d ed. (London: Barrie and Jenkins).

Douglas, M., 1975, *Implicit Meanings: Essays in Anthropology* (London: Routledge and Kegan Paul).

Ember C. R., 1977, 'Cross-Cultural Cognitive Studies,' in Siegel *et al.* (eds.), *Annual Review of Anthropology* (Palo Alto: Annual Reviews, Inc.), Vol. 6, pp. 35–56.

Feleppa, R., 1982, 'Translation as Rule-Governed Behavior,' *Philosophy of the Social Sciences* **12**, pp. 1–31.

Feleppa, R., 1986, 'Emics, Etics, and Social Objectivity,' *Current Anthroplogy* **27** (1986), pp. 243–255.

Feleppa, R., 1988a, *Convention, Translation, and Understanding* (Albany: State University of New York Press).

Feleppa, R., 1988b, 'Physicalism, Indeterminacy. and Interpretive Science,' *Metaphilosophy*, forthcoming.

Fisher, L. E., and O. Werner, 1978, 'Explaining Explanation: Tension in American Anthropology,' *Journal of Anthropological Research* **34**, pp. 194–218.

Frake, C. O., 1962, 'The Ethnographic Study of Cognitive Systems,' in Gladwin T. and W. C. Sturdevant, *Anthropology and Human Behavior* (Anthropological Society of Washington, Washington, D.C.), pp. 72–93.

Frake, C. O., 1964 'Notes on Queries in Ethnography,' in Romney, A. K. and R. G. D'Andrade (eds.) *Transcultural Studies in Cognition, American Anthropologist*, Vol. 66, No. 3, Pt. 2, Special Publication), pp. 132–145.

Frake, C. O., 1977, 'Plying Frames Can Be Dangerous: Some Reflections on Methodology in Cognitive Anthropology,' *Quarterly Newsletter of the Institute for Comparative Human Development*, Rockefeller University, Vol. 1 No. 3, pp. 1–7.

Geertz, C., 1973, *The Interpretation of Cultures* (New York: Basic Books).

Geertz, C., 1983, *Local Knowledge: Further Essays in Interpretive Anthropology* (New York: Basic Books).

Geoghegan, W. H., 1969, *Decision-Making and Residence on Tagtabon Island*, Working Paper No. 17, Language-Behavior Research Laboratory, University of California, Berkeley.

Gladwin, C., 1975, 'A Model of the Supply of Smoked Fish from Cape Coast to Kumasi,' in Plattner, S. (ed.), *Formal Methods in Economic Anthropology*, Special Publication No. 4 (Washington, DC: American Anthropological Association); pp. 77–127.

Goodenough, W., 1951, *Property, Kin, and Community on Truk*, Yale University Publications in Anthropology, No. 46 (New Haven CN: Yale University Press).

Goodenough, W., 1956, 'Componential Analysis and the Study of Meaning,' *Language* 33 (1956), pp. 195–216.

Goodenough, W., 1964, 'Cultural Anthropology and Linguistics,' in Hymes, D. (ed.), *Language in Culture and Society* (New York: Harper & Row), pp. 36–39.

Goodenough, W., 1970, *Description and Comparison in Cultural Anthropology* (Chicago: Aldine).

Goodman, N., 1968, *Languages of Art* (Indianapolis: Bobbs-Merrill).

Goodman, N., 1972, 'The Way the World Is,' in *Problems and Projects* (Indianapolis: Bobbs-Merrill), pp. 24–32.

Goodman, N., 1973, *Fact, Fiction, and Forecast*, 3rd ed. (Indianapolis: Bobbs-Merrill).

Goodman, N., 1977, *The Structure of Appearance*, 3d ed. (Dordrecht: Reidel).

Goodman, N., 1978, *Ways of Worldmaking* (Indianapolis: Hackett).

Goodman, N., 1984, *Of Mind and Other Matters* (Cambridge: Harvard).

Gumperz, J. and D. Hymes, 1972, *Directions in Sociolinguistics: The Ethnography of Communication* (New York: Holt, Rhinehart, and Winston).

Harris, M., 1968, *The Rise of Anthropological Theory* (New York: Crowell).

Harris, M., 1976, 'History and Significance of the Emic/Etic Distinction,' in Siegel *et al.* (eds.), *Annual Review of Anthropology* (Palo Alto: Annual Reviews, Inc.), Vol. 5, pp. 329–350.

Harris, M., 1979, *Cultural Materialism: The Struggle for a Science of Culture* (New York: Random House).

Hart, H. L. A., 1961, *The Concept of Law* (Oxford: Oxford).

Hollis, M., 1970a, 'The Limits of Irrationality,' in Wilson, B. (ed.), *Rationality* (Oxford: Blackwell), pp. 214–220.

Hollis, M., 1970b, 'Reason and Ritual' *loc. cit.*, pp. 221–239.
Hollis, M., 1979, 'The Epistemological Unity of Mankind,' in Brown, S. C. (ed.), *Philosophical Disputes in the Social Sciences* (Brighton: Harvester), pp. 225–232.
Hymes, D. H., 1970, 'Linguistic Method in Ethnography: Its Development in the United States,' in P. L. Garvin (ed.), *Method and Theory in Linguistics* (The Hague: Mouton), pp. 249–326.
Jorion, P., 1983, 'Emic and Etic: Two Anthropological Ways of Spilling Ink,' *Cambridge Anthropology* 8, pp. 41–68.
Kay, P., 1970, 'Some Theoretical Implications of Ethnographic Semantics,' *Current Directions in Anthropology*, American Anthropological Assoc. Bulletin, Vol. 3, No. 3, Pt. 2.
Kiefer, C., 1977, 'Psychological Anthropology,' in Siegel *et al.* (eds.), *Annual Review of Anthropology* (Palo Alto: Annual Reviews, Inc.), Vol. 6, pp. 103–119.
Kripke, S., 1982, *Wittgenstein on Rules and Private Language* (Cambridge: Harvard).
Larson, D., 87, 'Correspondence and the Third Dogma,' *Dialectica* 41, pp. 231–237.
Lear, J., 1982, 'Leaving the World Alone,' *The Journal of Philosophy* 79, pp. 382–403.
Lerner, D. (ed.), 1961, *Quantity and Quality* (New York: Free Press).
Lévi-Strauss, C., 1969, *The Raw and the Cooked: Introduction to a Science of Mythology*, trans. J. and D. Weightman (New York: Harper & Row).
Marano, L., 1982, 'Windigo Psychosis: The Anatomy of an Emic-Etic Confusion,' *Current Anthroplogy* 23, pp. 385–412.
Pike, N., 1954, *Language in Relation to a Unified Theory of the Structure of Human Behavior: Part I* (Glendale, CA: Summer Institute of Linguistics).
Pike, N., 1956, 'Towards a Theory of the Structure of Human Behavior,' in *Estudios Antropologicos Publicados en Homenaje al Doctor Manuel Gamio* (Mexico, D.F.: Sociedad Mexicana de Antropologia), pp. 659–671.
Quine, W. V. O., 1951a, 'Two Dogmas of Empiricism,' in his *From a Logical Point of View*, 2d ed. (Harper and Row, 1961), pp. 20–46.
Quine, W. V. O., 1951b, 'The Problem of Meaning in Linguistics,' *loc. cit.*, pp. 47–64.
Quine, W. V. O., 1960, *Word and Object* (Cambridge: M.I.T.).
Quine, W. V. O., 1968, 'Epistemology Naturalized.' In *Ontological Relativity and Other Essays* (New York: Columbia), pp. 69–90.
Quine, W. V. W., 1970, 'On the Reasons for Indeterminacy of Translation,' *The Journal of Philosophy* 67, pp. 179–183.
Quine, W. V. O., 1979, 'Facts of the Matter,' in Shahan, R. W. and C. Swoyer, *Essays on the Philosophy of W. V. Quine* (Norman: University of Oklahoma).
Quine, W. V. O., 1981, 'Goodman's *Ways of Worldmaking*,' in *Theories and Things* (Cambridge: Harvard), pp. 96–99.
Quine, W. V. O., 1987a, 'Indeterminacy of Translation Again,' *The Journal of Philosophy* 84, pp. 5–10.
Quine, W. V. O., 1987b, 'Reply to Paul A. Roth,' in Hahn, L. and P. Schilpp (eds.), *The Philosophy of W. V. Quine* (Open Court Press, 1987), pp. 459–461.
Quinn, N., 1978, 'Do Mfantse Fish Sellers Estimate Probabilities in Their Heads?' *American Ethnologist* 5, pp. 206–226.

Rawls, J., 1971, *A Theory of Justice* (Cambridge: Harvard).
Rorty, R., 1982, 'Method, Social Science, and Social Hope,' in *Consequences of Pragmatism* (Minneapolis: University of Minnesota), pp. 191–210.
Rudner, R., 1973, 'Some Essays at Objectivity,' *Philosophical Exchange* 1, pp. 115–135.
Schutz, N., 1975, 'On the Anatomy and Comparability of Linguistics and Ethnographic Description: Toward a Generative Theory of Ethnography,' in Kinkade, M. D. *et al.* (eds.), *Linguistics and Anthropology: In Honor of C. F. Voegeling* (Lisse: De Ridder), pp. 531–596.
Strawson, P., 1959, *Individuals* (London: Methuen).
Triandis, H. C., 1976, 'Approaches Toward Minimizing Translation,' in Brislin, R. W. (ed.), *Translation: Application and Research* (New York: Wiley), pp. 229–244.
Watson, L. C., 1981, '"Etic" and "Emic" Perspectives on Guajiro Urbanization,' *Urban Life* 9, pp. 441–468.
Werner, O., 1972, 'Ethnoscience 1972,' in Siegel *et al.* (eds.), *Annual Reviews in Anthropology* (Palo Alto: Annual Reviews, Inc.), Vol. 1, pp. 271–308.
Winch, P., 1958, *The Idea of a Social Science and Its Relation to Philosophy* (London: Routledge and Kegan Paul).

JOSEPH MARGOLIS

MONISTIC AND DUALISTIC CANONS FOR THE
NATURAL AND HUMAN SCIENCES

I

In a somewhat oblique but economical way, a review of Thomas Kuhn's influential *The Structure of Scientific Revolutions*,[1] helps to identify the current state of the methodology of the social and human sciences. Kuhn himself, of course, is very much interested in applying his theory to these sciences, though he has yet to pursue the matter in a sustained way. He has attracted a good deal of sanguine interest among the practitioners of the human sciences, of course. But the point, frankly, of beginning with Kuhn (and of pursuing at some length the promise of his well-known theory) is to show that the effectiveness with which he breached the canonical picture of the physical sciences does not really depend on his own favored notions of paradigm shifts and incommensurability (which are in any case not at all strongly defended or fully defensible) but on certain subterranean themes, somewhat displaced by his own notions, that *do* force a review of the methodological features of the physical sciences – either to justify a functional division between them and the human sciences or to confirm the dependence of the first *on* the second in a respect the canon has traditionally resisted or ignored (at least since it assumed its characteristically modern form). Kuhn managed to force the champions of the canon – which, at its boldest, appears as the unity of science program – to acknowledge seriously the implications of human agency in science, historical existence, the contingency and at least partial discontinuity of conceptual schemes, the technical and praxical finitude of human efforts at understanding nature, the relative opacity of nature *vis-à-vis* cognition, the absence of foundational access to the structures of the real world. By an irony of history, he accomplished this by a sort of indirection – which acquired a distinctly independent life. In effect, he 'annoyed' the establishment from within its own ranks by frontally attacking the received view: at the same time, deeper objections embedded in his own doctrine exposed the weakness of the canon *and* of his own emendation. There is a malicious complexity in the inadvertence of all of this. But it is worth pursuing because, now, the human sciences

155

Barry Glassner and Jonathan D. Moreno (eds.),
The Qualitative-Quantitative Distinction in the Social Sciences. 155–178.
© 1989 *by Kluwer Academic Publishers.*

need no longer be thought to be *dependent in a principled way* on the superior, adequate canon of the physical sciences. Also, from this vantage, it proves to be very easy to survey the principal lines of development regarding the social and human sciences as such. We should be able, then, to isolate the most important instructions for the emancipated pursuit of those studies.

One of the intriguing issues featured in Kuhn's book is fixed by the following not unreasonable remark:

> ... the historian of science may be tempted to exclaim that when paradigms change, the world itself changes with them. ... Nevertheless, paradigm changes do cause scientists to see the world of their research-engagement differently. ... The world that the student then enters [that is, as a result of changes in what inventive scientists come to see as a result of revolutionary transformations] is not, however, fixed once and for all by the nature of the environment, on the one hand, and of science, on the other. Rather, it is determined jointly by the environment and the particular normal-scientific tradition that the student has been trained to pursue. Therefore, at times of revolution, when the normal-scientific tradition changes, the scientist's perception of his environment must be re-educated – in some familiar situations he must learn to see a new gestalt.[2]

Kuhn never explicitly decided whether paradigm shifts entail different worlds or changes in our perception of an otherwise unitary world. He has been tempted by both possibilities; and here at least, he opts for a compromise that fails to address the issue squarely. No one would now deny that if perception is a real phenomenon, causally affected by paradigm changes, then *the world* must change as a result of changes in our way of perceiving. But it is quite another matter to hold that the world we inhabit perceptually is numerically dfferent from those others inhabit if their perception is paradigm-indexed in another way than ours. The dispute has been exacerbated, in analytic philosophical circles, by the somewhat irritated disagreement between W. V. Quine and Nelson Goodman – both of whom Kuhn has found suggestive – who take diametrically opposed views of the matter (without discussing paradigms); as well as in certain Continentally influenced quarters, for example, among so-called Yale deconstructionists, who, like Paul de Man, erase the demarcation between the real world and fictional worlds (since, on an extreme Nietzschean reading, we are said to inhabit language, the worlds of our fictions, even worlds constructed in accord with our 'metaphoric lies' about 'the' world). On that extreme thesis, there is *no* prospect that we can inhabit one real, actual, and all-encompassing world.

The fact is that talk of one actual world or of multiple worlds is a peculiarly inclusive way of organizing the objects of our discourse, not a claim that any merely sensory experience, for instance, could possibly support or weaken. It is notorious that even those who hold – Goodman for example[3] – that we inhabit different worlds because of our different conceptions-of-the-world, have not quite faced up to the fact that *we* seem able to entertain those alternative conceptions, that it is difficult to suppose that we ourselves discontinuously inhabit different worlds if, *via* a Kuhnian revolution, *we* make a paradigm shift. On the other hand, it is not really clear that those who hold that we do inhabit different worlds mean to deny (in so saying) that there is but one actual world; nor is it clear whether they could say straight out what the relationship might be, if any, between such paradigm-indexed 'worlds' and 'the' world.

The issue is not altogether verbal, however. One question of importance that hangs on it is this: on the paradigm-indexed-worlds view, truth (indistinguishable now from truth claims) cannot but be relativized to the world thus formed, and on the one-world view, the concept of truth *may* be reserved for some favorable relationship (however inchoate) between assertions and the world, while the *appraisal* of truth *claims* may be relativized to admitted evidence, theory choice, and paradigm allegiance (that is, treating talk of multiple worlds as signalling only a procedural restriction on an operational or idealist or instrumentalist management of truth claims) – without affecting or preempting the independent function of the theory of truth. This distinction can be maintained even if one denies a defensible demarcation between scientific realism and scientific idealism.[4] But the many-worlds view cannot coherently provide for a conception of truth (distinguished from a tally of the operative criteria for favorably appraising claims), timelessly or uniformly suited to any putatively alternative 'world.' (Here, it should be said: realism signifies a world accessible to human cognition but not merely an artifact of conceptual construction; and idealism signifies that, whatever the objects of human cognition, we can identify the structures and properties of none in any way constitutively independent of the categories by means of which they are humanly recognized. The import of this distinction becomes more troublesome when we apply it to the phenomena of the human sciences.)

There is a most important corollary question affecting Kuhn's notion of a paradigm. Kuhn has considerably modified his originally robust concept. He now no longer treats paradigms as straightforwardly and disjunctively enumerable – in terms of scientific communities, shared exemplars, 'disciplinary matrices,' sets of values and beliefs, or boundary conditions of any sort – not even in terms of the time it takes to form a paradigm or the time it takes to shift from one to another.[5] Hence, to the extent he is tempted to treat 'worlds' as paradigm-indexed, he can only admit a correspondingly fuzzy notion of the difference between the worlds of the partisans of different paradigms and even of the necessity for any paradigm being in place. But Kuhn also claims that he is 'a convinced believer in scientific progress'[6] – which would be quite meaningless if he held to a many-worlds view as opposed to the one-world view, and which, in any case, is not obviously reconcilable with his own notion of discontinuous paradigm shifts (unless in the Popperian way he apparently wishes to avoid).

In context, it is clear that, without having resolved the matter, Kuhn is primarily concerned to move against the realist/idealist disjunction: "There is, I think [he remarks], no theory-independent way to reconstruct phrases like 'really there'"; *and* to expose the bankruptcy of the competing view of scientific progress championed by Karl Popper: 'One often hears that successive theories grow ever closer to, or approximate more and more closely to, the truth.'[7] Popper *cannot* explain the bare eligibility of speaking of scientific progress under the condition of revolutionary science (in Kuhnian terms – in terms of paradigm shifts); although it was the profoundly conservative purpose of Imre Lakatos's conception of research programs to redeem (under alteration) Popper's original vision,[8] and although Popper himself is hardly obliged to concede that *he* has no consistent theory of scientific revolution.[9] Kuhn also, however, seems unable to explain the coherent connection between his own optimism about scientific progress and his insistence on discontinuous paradigm shifts that disallow a *measure* of progress from shift to shift. And indeed, there is no principled way of recovering progress under Kuhn's conditions.

Kuhn has now very significantly transferred much of what he first assigned to the presence or absence of a numerically distinct paradigm to the variable development of a putatively usable paradigm. Thus, he says:

the transition [to the maturity of a science] need not (I now think should not) be associated with the first acquisition of a paradigm .... What changes with the transition to maturity is not the presence of a paradigm but rather its nature. ... Many of the attributes of a developed science which I have above associated with the acquisition of a paradigm I would therefore now discuss as consequences of the acquisition of the sort of paradigm that identifies challenging puzzles, supplies clues to their solution, and guarantees that the truly clever practitioner will succeed. Only those who have taken courage from observing that their own field (or school) has paradigms are likely to feel that something important is sacrificed by the change.[10]

The reason this seemingly modest modification is of the utmost importance is that the very meaning of Kuhn's famous claims about incommensurability is decisively affected by the alteration. Notice that Kuhn does not abandon the notion of a paradigm – or even concede that, in the 'usual' practice of science (meaning by that term to avoid the dispute about 'normal' and 'extraordinary' or 'revolutionary' science), paradigms may not obtain at all; or we may not be able to tell whether a paradigm has simply evolved or replaced another or even been present in the first place. Popper for instance holds that, regarding the history of the theory of matter since antiquity, the Kuhnian account does not fit, as it does not, also, in the biological sciences; it may, he says, 'fit astronomy fairly well.'[11] But it is quite impossible to confirm that anything like Kuhn's notion of normal science obtains in what are variously called the psychological, behavioral, social, historical, linguistic, cultural, or (generally) human sciences. Kuhn is tempted to convert his 'descriptive' account into a 'normative' one – or at least to favor a theory that indissolubly entails both considerations.[12] But the very plausibility of such a move is one of the decisive issues at stake. Also, Kuhn's blitheness here betrays the sense in which *he* means to provide a (radically) modified account of the unity of science program – still governed by the model of physics and astronomy – when the upshot of his own intervention contributes to dismantling that very canon. Kuhn himself, it should also be remarked, notes that the change in the concept of the maturity of a science is particularly pertinent to assessing the work and expectations of 'those concerned with the development of the contemporary social sciences.'[13] So the issue is joined.

Now, it is no longer clear – given these adjustments – whether, as Kuhn himself has always maintained, his strongest critics have simply misunderstood him. Correspondingly, it is no longer clear what the order of adequacy is among the various theories of science ('paradigmatically,' the physical sciences) that are caught up in the quarrel between Kuhn and the Popperians (and others); *and*, furthermore, *it is no longer clear what the conceptual relationship is or should be construed as being between the natural and the human sciences.* This is what has been radically obscured and radically illuminated by that extremely murky dispute. Very roughly put, the bearing of Kuhn's original contribution on the ontology and methodology appropriate to the human sciences rests with the fortunes of his incommensurability thesis; *that* thesis depends essentially – though hardly exclusively – on the doctrine of paradigms and paradigm shifts and on the alleged relationship between paradigms and 'the' world or the plural 'worlds' we 'inhabit.'

But *those* doctrines are both displacements from the real issue and ultimately untenable.

There are really two parts to Kuhn's incommensurability thesis: one concerns the problem of translation, particularly among paradigm-indexed statements; the other concerns the problem of theory-choice where competing theories are said to be paradigm-indexed. Kuhn's views on both surely have a modicum of truth; but neither of them – not even the two jointly – has the full power they were first alleged to have. If anything, the improved theory of paradigms cannot but weaken the incommensurability thesis. And if that is so, then Kuhn's opponents cannot rightly be supposed to have simply misunderstood him (in conflating intelligibility with commensurability – across putative paradigms); although conceding that does *not* yet vindicate the theory of any of Kuhn's opponents any more than his own, and although the ulterior import of this provisional stalemate *does* directly affect in a most important (and unperceived) way the methodological standing of the human sciences. The entire tradition of analytic philosophy of science, from the positivists (even of the latter half of the nineteenth century) through the Vienna Circle and the unity of science program, through Popper and Lakatos and Feyerabend, has been committed to the suitability of theorizing about the natural and human sciences *together*. This is true (perhaps surprisingly) even of Kuhn, even though Kuhn's challenge to what may be called the 'uniformitarian' thesis[14] (that is, his challenge to the denial of an essential distinction between normal and revolutionary science) – *does* (unintentionally) reopen the question of the relationship between the natural and human sciences, radicalized in a way because of the serious weaknesses uncovered in the rival theories *and* in his own, by his own original work. This, at any rate, sketches the strategic reason for having invoked Kuhn as we have – and for having pursued certain details in the extended way we have.

The incommensurability thesis suffers from a number of general difficulties, which are not entirely of Kuhn's own making. For one thing, there is *no* well-received working theory for accurately fixing the actual meanings of terms so that *shifts* or *incommensurabilities* among the meanings of given terms – in the very strong sense Kuhn and Feyerabend favor – could even be reliably decided.[15] In fact, on the apparent thesis, there *could* be no such working theory, since the operative criteria would, *per impossibile*, be neutral to paradigm (or theory) shifts. More important, Kuhn's most careful formulation of what incommensurability comes to, in comparing the terms of two successive theories, proves to be a ubiquitous feature of all natural languages; in fact, Kuhn himself favorably cites Quine's well-known indeterminacy of translation thesis and claims (misleadingly) to have 'extended' Quine's account (to apply apparently to predicates or to expressions making predications of this and that).[16]

The actual claim is this: that there is at hand no suitable vocabulary (and none that could be developed) to preclude incommensurability; 'The point-by-point comparison of two successive theories [Kuhn says] demands a language into which at least the empirical consequences of both can be translated without loss or change,'[17] and that is not possible. But first, since this is the general condition of natural language (at least on Kuhn's own view, on Quine's on Feyerabend's – which Kuhn takes to be similar to his own – and on others', at least by default), the incommensurability theory *does not depend on, require, or specifically illuminate the notion of a paradigm or paradigm shift*; and second, in spite of such incommensurability, natural language speakers – one must suppose, the partisans of different paradigms as well – *can understand one another and make suitable translations of one another's discourse* (though hardly 'without loss or change,' a distinction now no longer even clearly measurable). The failure to provide criteria for individuating paradigms, the absence of clear boundary conditions for paradigms, the late change in the theory of paradigms, all conspire to weaken the importance of the charge of incommensurability. This is why the objections of Kuhn's opponents are not, as Kuhn is disposed to say, simply confusing intelligibility and commensurability.[18] Small incommensurabilities seem to obtain everywhere in a living language; but that 'fact' (conceded at face value) is itself actually spotted linguistically *and* does not appear to affect intelligibility, translatability, or even the provision of specialized rules for insuring commensurability wherever needed or wanted. Discarding the strongly disjunctive view of paradigms eliminates as well the sole foundation for a strongly pertinent claim of incommensurability – at least as far as translatability is concerned.

The crucial impact of Kuhn's thesis lies not with the theory of paradigms but rather with the import: (1) *of denying that there is or could be a neutral language into which competing or successive theories may be translated*; (2) *of denying that reality is cognitively transparent and indifferent (anywhere) to our conceptual schemes*; (3) *of insisting that conceptual changes have a nonlinear, discontinuous, openended, unpredictable, and nontotalizable character*. These conditions, however, are now generally acknowledged through the whole of Western philosophy and specialized sciences, with or without regard to the theory of paradigms or to the one-world/many-worlds quarrel. Their theoretical importance, somewhat displaced by the fashionable claims of paradigm shifts and incommensurability, lies rather with the exposed dependence of the natural sciences themselves (construed as a form of cognitive activity) *on* the very conditions of human existence – that is, on the analysis of the phenomena first examined by those 'weaker' would-be sciences that the unity of science program had always insisted must come to heel, to deserve the name of science in whatever measure can be justified. So the tables have been turned.

Our inability to formulate a principle or rule or rational method or policy for *systematically and conclusively* validating choices among theories indexed to different paradigms is simply a special case (if it obtains at all) of our general inability to formulate the once-and-for-all assured grounds for such high-level choices – *for all shifts in the methodology of science or rational inquiry.* It is no wonder we lack the ability. But (once again) that signifies only that nature is not cognitively transparent, that human inquiry is not reliably progressive in approximating the ultimate structure of an independent world, that our educated guesses about how best to proceed within any inquiry are affected by contingent inquiry and by our ignorance of the relationship between what we may know and what we clearly do not yet know. Furthermore, *this* inability does *not* support the claim of incommensurability (*or* of relativism, with which it is wrongly identified).[19] It is not incommensurability that is at stake; it is only *consistency* with our (and Kuhn's) having rejected such doctrines as that of the linguistic or conceptual neutrality of science, of the cognitive privilege of favored forms of inquiry, of the assurance that we are approximating ever more closely to the 'truth' about the world, of the benign contingency of our historical existence (as not affecting our ability to discern the essential structures of nature), and the like.[20]

The master importance of Kuhn's contribution lies here. It lies with having introduced in the most *infectious* way possible, at the very heart of the tradition of the philosophy of science, a variety of insights collected from various sources (collected by others from such disparate and unlikely figures as Dewey and Heidegger, for instance)[21] that are simply wrongly presented (by Kuhn himself) as having to do primarily with the significance of paradigms and the cognate problem of incommensurability. That peculiar form of *masking* the real import of his work has obscured *both* the tenuous status of current philosophies of the natural sciences *and* its true bearing on the fortunes of the human sciences – affected by what Kuhn has successfully uncovered. The point may be put this way: Kuhn has shown (not altogether in accord with his own intention) that the strongest *physical* sciences are radically affected by the historical contingencies of the methods of *human* inquiry – evolving without any evidential assurance of their being able to discover progressively the underlying structures of nature. What this means, in effect, is that *the peculiar features of the phenomena the human sciences specifically examine inextricably color the study of the phenomena of the natural sciences; although what the properties of the former are, what the methodological features of the study of the former are best construed as being, what the conceptual relationship is between the two sorts of science, whether there is a unified method for studying phenomena of both kinds, appear to be very nearly completely unexamined from the vantage of the tradition of theorizing about the physical sciences leading to and immediately beyond Kuhn's own intervention.*

Kuhn's account does not confirm or fix the reasons for the present crisis regarding the epistemic and methodological foundations of the physical sciences; it offers, instead, a speculation about the sociological causes of that crisis mistakenly cast as a conceptual or logical or philosophical analysis of the very properties of scientific inquiry. Its rather amazing meteoric career is largely due to the fact that it has managed, in an initially plausible way, to convey what *are* very probably the main ingredients of the present conceptual crisis, distorted and displaced so as to strengthen the sense in which theorizing about the physical and the human sciences *together* appears to continue the master tradition of the philosophy of science itself. But the genuinely subversive philosophical import of Kuhn's account lies in his having successfully *confronted that tradition with the need to reconcile its principal claims with the import of the thoroughly historicized, cognitively non-privileged nature of human inquiry, addressed to a world perceived in accord with diachronically changing concepts, imperfectly known to those who use them thus.* Once we view matters this way, it is comparatively easy to sketch an ideal declension of the recent history of the philosophy of science down to its present impasse and, in doing that, to identify as well the remarkably virgin territory now stretching before the human sciences.

II

The history of the philosophy of science is largely marked by the gradual abandonment of just those doctrines the conceptual discrediting of which (established on independent grounds) is precisely what Kuhn manages to convey in his attack on recent positivist and Popperian views. The force of the argument does not lie in Kuhn's actual attack but rather in the inherent inadequacy of those theories, somehow magnified by Kuhn's 'distortion' – which has the further peculiar force of exposing the inadequacy of his own attempt at an emended theory of science *and* of confirming the reasonableness of *freeing the methodology of the human sciences from the rather muddled methodology of the physical sciences*. Admittedly, this is a bold claim that, though it may deserve a hearing, requires a defense. We can afford here only a sketch of the salient phases of that history.

The most sanguine, most extreme, most reductively committed, most systematic, most comprehensive, and most characteristically oriented short picture of the language and method of a 'genuine' science, formulated in our century (in the thirties) is probably adumbrated in the following statement by Rudolf Carnap – worked out at about the high point of the collaboration (and developing divergence) between Carnap and Moritz Schlick, at the time the Vienna Circle was the leading center for the unity of science program:

> To every sentence of the system language [psychology, here] there corresponds some sentence of the physical language such that the two sentences are inter-translatable. ... The various protocol languages [reports of sensory experience and the like are] sublanguages of the physical language. *The physical language is universal and inter-subjective* [physicalism]. ... [T]he generalized sentences of psychology, the *laws* of psychology, are also translatable into the physical language. They are thus physical laws. Whether or not these physical laws are deducible from those holding in inorganic physics, remains, however, an open question. This question of the deducibility of the laws is completely independent of the question of the definability of concepts.[22]

Carnap of course took his physicalism to extend to the social sciences.[23]

The unity of science program has tended to diverge in a variety of ways from Carnap's very strong thesis. By and large, however, it has emphasized the somewhat different (but related) objectives of (a) the unity of scientific language (in its most extreme form, Carnap's physicalism), (b) the unity of the laws of all sciences (produced, as in the well-known account by Paul Oppenheim and Hilary Putnam, by successive 'microreductions,' or by deriving all such laws from 'the laws of some one discipline'[24]), (c) the unity of the laws of the reducing discipline (linking microreduction or analogous strategies to some set of suitable part/whole conceptions of the elements of all the domains involved)[25], and, finally (d) the unity of the explanatory method of science itself (most famously formulated in the so-called deductive-nomological model of explanation).[26]

Apart from technical difficulties in the various versions of the program,[27] the entire undertaking may be challenged in essential ways by the following quite elementary considerations. First of all, the original positivist vision was committed to a form of foundationalism with regard to the protocol sentences on which the entire program allegedly rested – which doctrine, even within the Vienna Circle, Otto Neurath had succeeded in completely demolishing (an achievement acknowledged by both Carnap and Schlick).[28] Foundationalism has never been seriously reaffirmed in any form within the canonical view of the physical sciences; but its rejection entails at the very least that we cannot specify universally or timelessly reliable sources from which all would-be sciences proceed and in accord with which the actual structure and properties of the real world are known to be directly discriminable.[29] There must be a concession here to unspecified conditions of epistemic opacity and to the unknown historical contingencies of scientific inquiry. In the hands of a strong critic of positivism like Popper, it then became extremely easy to hold that the very pretense of having *discovered*, by means of empirical science, the actual laws of nature constituted an utterly indefensible commitment to essentialism, that is, to the cognitive transparency of the invariant structure of nature.[30] But of course the retreat from foundationalism and essentialism could not but adversely affect the presumption of both the adequacy of physicalism *and* of the very possibility of sustaining a program of progressive microreductions and deductive-nomological explanations. These prospects were rendered inherently uncertain.

Secondly, there is no optimistic clue following the analysis or reduction of actual linguistic phenomena and behavior in *sub*-linguistic terms. Language appears to be *sui generis* in a very strong sense.[31] Without overcoming its apparent irreducibility, there would be no basis for supposing that the language and method of the human sciences *could* in principle be adequately characterized in terms of the language and method alleged to be adequate for the physical sciences. Some principled bifurcation of the sciences would become inevitable.

Thirdly, once physicalism and foundationalism were denied, we should have to face the near certainty that the regularities of the human sciences could not be construed as or as approximating universal covering laws. The issue is not simply one of substituting statistical laws for laws that explicitly express a universalized invariance; the very idea of an invariant limit ordering statistical regularities among cultural phenomena would appear completely arbitrary, possibly even a prejudice designed to shore up the deductive-nomological model of explanation.[32] Given the bifurcation of linguistic and nonlinguistic phenomena – hence, the failure of progressive microreduction in the social and cultural disciplines – it would become increasingly unclear why it should be supposed that, even in principle, universal nomic invariances obtain among linguistically qualified phenomena (for instance, in political history and in the fine arts).

Fourthly, the peculiar properties of human language (taken now as emergent and *sui generis*) appear to be such that the very concept of causality would be affected by the admission that linguistically informed agents do, as such, play a distinct causal role. In particular, if the intensional features of language could not be satisfactorily regimented in accord with so-called extensionalist strategies (for instance, as proposed by Quine and Donald Davidson),[33] then we should have to admit that causal contexts were not invariably extensional and that causal explanation could not invariably involve covering laws; in fact, causality and nomologicality would be seen to be independent notions.[34] This, of course, would provide a very strong basis for favoring the methodological distinction and independence of the human sciences.

Finally, if foundationalism, essentialism, and related doctrines were indefensible, then the realism of science could not be formulated so as to insure a strong demarcation between scientific realism and scientific idealism, that is, could not be formulated without conceding that all inquiry is inextricably conditioned by conceptual categories native to and acquired by human investigators. Once that admission was construed in terms of the contingencies of historical existence, the profound conceptual dependence of the physical sciences themselves *on the human sciences* would become entirely evident.[35] In fact, these objections to the unity of science program pretty well capture the fundamental themes of Kuhn's somewhat displaced criticism of the canonical picture.

Having come this far, we can, now without fear of misunderstanding, say that methodological theories of the human sciences have tended to be strongly polarized as monistic or dualistic; that the monistic views (essentially, versions of the unity of science) have been obliged to construe more and more generously the complexity of phenomena falling within the 'physical' or 'natural' sciences; and that the dualistic views (perhaps first formulated, for modern discussions, by Wilhelm Dilthey) have been obliged to construe the bifurcation of the sciences less and less in terms of a Cartesian ontology (Dilthey's original bias) and more and more in terms of a distinctive basic vocabulary, form of cognitive access, and the analysis of causality. The notion of unifying the sciences is probably irresistible – and entirely reasonable. But it must be appreciated that any unification based on the methodological nonreducibility of the human sciences to the physical *and* based on the conceptual dependence of the physical sciences themselves on the human sciences is utterly inimical to the original objective of the unity of science program.

Without pretending to any simple historical lineage among the principal theories of the human sciences, we may schematize the strongest recent currents in the following way: (1) monistic views have explored the possibility of *enlarging* the scientific canon along structuralist lines – particularly, as in the work of Claude Lévi-Strauss, Jean Piaget, and Noam Chomsky; and (2) dualistic views have explored the possibility of reinterpreting the scientific canon as incorporating the physical sciences *within* a hermeneutic or praxical orientation – particularly, as in the work of the Marxists, the Frankfurt Critical theorists, the Freudians, and such allied thinkers as Paul Ricoeur, Michel Foucault, Jacques Lacan.

Se seen, the subversive distinction of Kuhn's work lies – despite his having intended to liberalize the monistic canon beyond positivist and Popperian constraints – in his having (by his own deflecting attack) actually exposed the impossibility of merely emending that canon as well as having contributed aid and comfort (inadvertently) to an improved dualistic view. The most obvious evidence of this is linked to his well-known theory of the discontinuity of paradigm shifts (the doctrine he keeps retreating from, in the interest of hewing as closely as possible to the unity of science model): the fact is that, in advancing that notion, Kuhn managed to expose (as did, also, Feyerabend and, inadvertently again, Lakatos) the essential arbitrariness of Popper's well-known (somewhat 'desperate') doctrine of verisimilitude (that is, of the progressive approximation to the actual structures of nature, without ever reaching them), once Popper *separated* the cognitive appraisal of verisimilitude from the cognitive appraisal of piecemeal scientific claims.[36] This is the crux of the decisive disordering of the monistic vision and its invasion by the dualistic view.

Popper (and in effect Charles Sanders Peirce before him, and inductivists like Hans Reichenbach at about the same time) fails to account for their generalized confidence in the *linear* progress of science – applied uniformly to all the sciences. Popper specifically denies (correctly, once having rejected essentialism) that there could be direct cognitive evidence of verisimilitude; but then he has (and can have) no grounds for construing the open society's use of falsifiability as favoring verisimilitude at all – in effect, he confuses scientific realism and verisimilitude.[37] Peirce openly embraces fallibilism with respect to scientific results, but he neglects to acknowledge that fallibilism applies reflexively to the *canons* of scientific inquiry itself; hence, his asymptotic theory of the match of truth and reality in the long run is completely undermined.[38] For his part, Reichenbach simply fails to explore the implications of an historicist conception of science for the viability of the inductivist confidence itself.[39] To put things in a somewhat lighthearted way, this is why Feyerabend's scientific anarchism is so difficult to dismiss: it is a kind of terribly knowledgeable blackmail against those who have simply stubbornly refused to draw the obvious consequences of the developing picture of the physical sciences.[40]

III

Turn, finally, to the current struggle between the monistic and dualistic conceptions of the natural and human sciences. A few specimens will have to do – and are actually sufficient for our purpose.

Three rather different kinds of structuralist sciences have been developed in recent years, each intended to provide a way of incorporating within the 'physical' or 'natural' sciences the distinctive phenomena of some sector of the 'human studies.' Chomsky, for one, generalizes about cognitive psychology on the basis of linguistic studies. His most distinctive commitment is perhaps the following: 'our minds are fixed biological systems with their intrinsic scope and limits.'[41] This he explicitly means to investigate in such a way that, as he says, 'the notion of 'physical body' [may need to] be extended ... to incorporate entities and principles of hitherto unrecognized character.'[42] The opposition to Carnap's reductionism is reasonably clear. But Chomsky means to construe the import of his own project as being in full accord with the rejection of 'the bifurcation thesis' – 'that is, the thesis that theories of meaning, language and much of psychology are faced with a problem of indeterminacy that is qualitatively different in some way from the underdetermination of theory by evidence in the natural sciences.'[43] The trouble is that *if* his well-known theory of the biologically native, species-specific deep grammar of human natural languages is false, *because* that putative grammar is *not* independent of (or altogether independent of) the contingent, historically shifting social experience of different, *sub*-species-wide communities, then the very advocacy of extending the monistic vision of science would be adversely affected as well. That is the decisive matter.[44] That is why, also, Chomsky says that if 'the connection between semantic representation and deep structure' cannot be idealized so as to insure the modular independence of deep structure from semantic representation – hence, also, to avoid incorporating 'nonlinguistic factors into grammar: beliefs, attitudes, etc. ... [then] I [that is, Chomsky] would conclude that language is a chaos that is not worth studying.'[45] In short, to admit the inseparability of grammar and meaning *and* the inseparability of that complex from the largely nonlinguistic phenomena of historical existence is to raise again the question of what it means to attempt to incorporate the human sciences into the *enlarged* canon of the natural sciences. *Certain changes among empirical hypotheses entail radical changes in the conception of science itself.*

Claude Lévi-Strauss offers what is perhaps the boldest and most uncompromising form of structuralism – certainly not nativist in Chomsky's sense, more clearly linked to the totalizing orientation of Saussure's analytic structuralism (that is, as socially but not infrapsychologically assigned) – but a variant still of the same systematic effort to extend the adequacy of what both view as the proper scope of the natural or physical sciences. So Lévi-Strauss says: 'I believe the ultimate goal of the human sciences to be not to constitute, but to dissolve man. ... Ethnographic analysis tries to arrive at invariants beyond the empirical diversity of human societies. ...'[46] Lévi-Strauss is rather vague (actually) about the status of the structuralist system in terms of which to 'place' ethnographic data; but it is instructive to note that (like Chomsky) he rejects the idea that the validity of the formal relations of his system (or grammar) is conceptually dependent on hermeneutic, semantic, or interpretive considerations. So he says, quite characteristically, specifically in opposition to Jean-Paul Sartre's historicism: 'thanks to anthropology itself, historical humanity has given the blessing of meaning to an original humanity which was without it ... a good deal of egocentricity and naivety is necessary to believe that man has taken refuge in a single one of the historical or geographical modes of his existence, when the truth about man resides in the system of their differences and common properties. ... Language, an unreflecting totalization, is human reason which has its reasons and of which man knows nothing.'[47] Lévi-Strauss nowhere accounts, however, for the circumstances of human inquiry in virtue of which *this* (or any) allegedly valid totalization – now, somehow prior to and independent of interpreting human events – can be both empirically confirmed and freed from the conceptual linkage between structure and meaning *and* the historical linkage between the development of an inquiring society and its understanding of the materials of any society observed (including its own). So once again, we cannot fail to see that some hypotheses bear in their wake radically different conceptions of what a science is: Lévi-Strauss appears to preclude precisely what would challenge the adequacy of extending the canon of the natural sciences into the human sciences.

Jean Piaget's developmental psychology is also a form of structuralism, one in fact in which the explanation of cognitive development among human children is said to be 'of the same kind' as what is required in accounting for 'the adaptation of an organism to its environment.' Basically, Piaget opposes the empiricist conception of 'a simple copy of external objects' and the nativist conception of 'a mere unfolding of structures preformed inside the subject'; his own view favors 'a set of structures progressively constructed by continuous interaction between the subject and the external world.'[48] A reasonably succinct summary of his view runs as follows:

> The problem of stages in developmental psychology is analogous to that of stages in embryogenesis. The question that arises in this field is also that of making allowance for both genetic preformation and of an eventual 'epigenesis' in the sense of construction by interactions between the genome and the environment. It is for this reason that Waddington introduces the concept of 'epigenetic system' and also a distinction between the genotype and the 'epigenotype.' The main characteristics of such an epigenetic development are not only the well-known and obvious ones of succession in sequential order and of progressive integration (segmentation followed by determination controlled by specific 'competence' and finally 'reintegration') but also some less obvious ones pointed out by Waddington. These are the existence of 'creodes,' or necessary developmental sequences, each with its own 'time tally,' or schedule, and the intervention of a sort of evolutionary regulation, or 'homeorhesis.' Homeorhesis acts in such a way that if an external influence causes the developing organism to deviate from one of its creodes, there ensues a homeoretical reaction, which tends to channel it back to the normal sequence or, if this fails, switches it to a new creode as similar as possible to the original one.[49]

So, like Chomsky and Lévi-Strauss, but by an altogether different route – avoiding either a fixed infrapsychological or a fully social system – Piaget identifies distinctive, invariant, interactional, structures that link in a necessary way all pertinent variations within a totalized system, *without any explicit attention at all to intentional or interpretive content and without any attention at all to the conditions of historicized existence that might alter or challenge in a crucial way our account of the uniformly species-specific development, behavior, or mentation of the human race.* The fact is that Lev Vygotsky had, from a point of view shaped by Marxist and allied considerations, already noted at a very early date Piaget's failure to say anything about the social nature and import of linguistic acquisition and the correspondingly socialized nature and import of distinctly human thinking.[50] With rather little adjustment, Piaget really never amended his view – which once again confirms the relationship between particular empirical hypotheses and the very concept of science or human science.

In these varieties of structuralism, then, we have what are perhaps the most ramified attempts to date to extend the *monistic* canon of science, without endorsing reductive accounts like those of Carnap or B. F. Skinner and without capitulating to dualism. But in resisting any and all versions of the dualistic orientation, these views characteristically treat history as a phenotypic variation of a genotype, or a contingent instantiation of an invariant and timeless rule, or a transient episode conforming with a fixed law of development or change.[51] Diachronic changes in interpretive understanding, constraints due to historical experience, praxical and technical achievements, shifts in semiotic orientation linked to these, the irreducibly intensional complexity of linguistic and linguistically qualified phenomena, the reflexive critique of our own changing effort to understand ourselves, are all essentially ignored – that is, precisely those themes insinuated into the dispute about canonical science by Kuhn's 'apostasy,' those very themes critical to the alleged distinction of the human sciences.

Probably the most convincing important instance of the grudging abandonment of the monistic view in favor of the dualistic – partly because it is unwelcome, inexplicit, baffling to the author himself; partly because it has been taken up by other more canny, more eclectic practitioners of the dualistic mode (Lacan and Habermas, for instance)[52] – appears in Freud's acknowledgement that the intended positivism of his own *Scientific Project* could not be sustained and that the method of psychoanalysis seemed more akin to the interpretation of literary narrative.[53]

Here, then, we have a glimpse of the changing picture. We cannot pretend to canvass more than a few possibilities. It is perhaps enough to identify the principal themes that mark the dualistic mode of construing the human sciences as not methodologically reducible to the natural sciences and (in fact) as imposing distinct constraints *on* the latter without denying (within the scope of such conditions) their different but equipotent undertakings. We may do this partly by illustration, but the master themes are already somewhat clear from what has been said. The important thing is to grasp the sense of contest between the two conceptions – not to press a premature resolution exclusively favoring either. So a simple list will serve our needs, confirming as is must a set of difficult and persistent questions. First, emphasis is placed, by advocates of the dualistic mode, on the historical existence of man, both as scientist and as object of inquiry. This is meant to bear on every cognitive endeavor – hence, second, to preclude any and all claims to escape the contingency of would-be knowledge, by way of foundations, privileged access, transparency of reality, essentialism, correspondence, self-evidence, systems totalized for all possible worlds, independently assured extensional reductions and universal regularities, and the like. Thirdly, our knowledge of the world, of any part of the world, is taken to be inextricably constrained by the categories of our own understanding, themselves subject to historical influence and change – hence, constrained in a way that disallows a clear demarcation between realism and idealism. Fourthly, within the contingencies of our finite history, and continuous with openended changes in conceptual orientation (that we realize are bound to come but that, thus limited, we cannot now fully fathom), we judge ourselves obliged to search who best to understand how our own present orientation may delude or mislead us or distort our picture of the world – that is, we attempt a critique of our mode of understanding within its own boundaries. Fifthly, we take ourselves and those who have preceded us to be or to have been causally efficacious agents, that their work and ours are intentionally qualified by the conceptual orientation of our respective societies – hence, qualified by intensional complexities of language and understood, only by way of historically, intensionally variable interpretations. Sixthly, we take it that the cognitive, conceptual, linguistic structure of every human society (that is, its cultural structure) cannot be completely internalized in any infra-psychological or infrabiological way, and accommodates diachronic changes in its own putative rules or practices not generable from those of a previous phase. Seventhly, language, which human history and human culture presuppose and which is itself the paradigm medium of both, is *sui generis*, that is, not analyzable in terms of any (or any known) *sub*-linguistic elements – also, at its own emergent level, apparently not a closed, finite, totalized system of any sort, however readily it may be idealized as such for any given *segment* of human practices. There are other distinctive themes associated with the dualistic mode, but these are probably the most central and the most difficult to neutralize in favor of any monistic conception of science. Admit these doctrines, and there is no prospect of retreating to the monistic view.

It remains only to afford a fair sense of how representative champions of the dualistic mode engage these themes (theorists of course who may well be attracted to a new unification of science, though not to the unity of science program itself). One thing they must do, however, to deserve full comparison with the advocates of the monistic view: they must reject anything like a Cartesian interpretation of the grounds for methodological dualism as well as anything like a merely Kantian or Hegelian acknowledgement of history. In effect, this means seriously qualifying at least Dilthey's division of the sciences. For, although Dilthey did attempt to match the rigor of the natural sciences in the work of the human sciences, his basic doctrine of inner experience (*Erlebnis*) is Cartesian in spite of his opposition to a presuppositionless but axiomatizable science.[54] To note that fact, however, cannot fail to suggest (or confirm) that there really are at present no fully ramified theories of the human sciences that escape both the disorders of confusing the would-be canonical objectives of the natural sciences with the legitimate objectives of the human sciences and the disorders of maintaining that difference at the price of adhering to vestigial philosophical doctrines ultimately incompatible with effectively sustaining the distinction – for instance, Cartesian certitude or Kantian transcendental necessity or Hegelian history or Husserlian *a priori* discipline. In effect, this marks the most promising instruction for those who would theorize about the human sciences.

Most currently salient views, sanguine still about the prospects of characterizing a genuine science, nevertheless manage to compromise that objective at a critical point. Hans-George Gadamer, for instance, who rather equivocally rejects the very notion of a hermeneutic method (since he also obviously contrasts valid or authentic interpretation and that which is merely gratuitous or uninformed), also manages to forestall the threatening relativism of his own account by falling back to a subtle version of essentialism somehow embedded and preserved through history itself. So he says, symptomatically acknowledging the 'primordial communality that unites all the thought attempts of humanity': 'each projection of universal history has a validity that does not last much longer than the appearance of a flash momentarily cutting across the darkness of the future as well as of the past as it gets lost in the ensuing twilight. That is the proposition of hermeneutical philosophy that I dared to defend against Hegel.'[55] Similarly, in the work of Jürgen Habermas, utterly contrary to the praxical orientation of his own blend of hermeneutic, Marxist, and Frankfurt Critical thinking, one must acknowledge a retreat to unsupportable universalist pretensions in Habermas's well-known account of the conditions of 'communicative competence.'[56] If, that is, human thought is genuinely constrained by the finite limits of historical and praxical existence, then timelessly valid universal norms of rationality (Habermas's objective) are neither cognitively required nor confirmable for science or successful communication. Marx himself might have been expected to offer a clear paradigm of the human sciences, except of course that the complexity of his thought baffles any prospect of a simple, single, and explicit model – as Alvin Gouldner's well-known account of the 'two Marxisms' amply attests.[57]

No, the best sense of the dangers to be avoided in formulating a fresh account of the human sciences, consistent with the dialectical difficulty of constructing *an historicized method for historicized phenomena*, is probably to be found in such disparate, uncompromising, and yet obviously optimistic statements like those of Theodor Adorno, Michel Foucault, Paul Ricoeur[58] – *not* in the allied but rather self-defeating, merely 'deconstructive' criticism of Jacques Derrida and Paul Feyerabend.[59] But the authors first mentioned do not yet provide a fully articulated methodological theory that would support a point-by-point comparison with the still impressive monistic views of the past. That is what we need and still lack. There is, moreover, no single model of the human sciences to be had; and there is no reason even to suppose that either in the natural or the human sciences one 'unified' conception of method could in principle collect (non-vacuously) all relevant inquiries. The best we are likely to be able to do at present is collect convergent features of the most promising overviews of the human sciences – which, merely by their being listed, confirm and fix the remarkable history that has led us from early positivism to our present sense of a new theoretical horizon.

These themes must surely include: (i) the hermeneutic import of science being itself the work of historically situated human agents; (ii) the capacity of such agents to exercise, within their own historical setting, the double ability of interpreting the work of those who have preceded them and of reflecting critically on the limitations of their own conceptual orientation; (iii) the sense in which the physical or natural world is investigated, ineluctably, within the enabling constraints of the conceptual categories of particular historical societies, despite the reasonableness of viewing its reality as not depending on such efforts of understanding; as well as the sense in which the reality of the human or cultural or linguistic or historical world is also independent of the lives of any of us taken singly, though hardly independent of the activity of entire societies that constitute that world and whose members examine and come to understand it; (iv) the significance of the fact that the same agents jointly pursue both kinds of science and that the world of human culture must manifest itself in the materials of physical nature; and (v) the complications that result from its being impossible, within the realtime constraints of historical existence, to know or assess correctly the full bearing of any present finite segment of human experience on the unknown, openended, future encounters on which and in accord with which our theory of nature and culture will depend and be adjusted.

To say, with Feyerabend and Gadamer for instance, that there can be no method in science is simply to have failed to appreciate the very sense in which their own reasonable criticism of antecedent prejudices requires a change in the conceptions of method and rationality themselves *and* in which we (and they) must suppose that we *can* characterize the process of human reason by which the episodic but reflexively continuous integration of the body of science is constantly reshaped. This is what we have to understand, and this is what, ultimately, we mean by the human sciences – applied either to their own distinctive fields of inquiry (the activity and work of human societies) or, hermeneutically still, to the activities of scientists addressed to the apparently independent phenomena of the natural sciences.

NOTES

[1] Thomas S. Kuhn, *The Structure of Scientific Revolutions*, 2nd ed. enl. (Chicago: University of Chicago Press, 1962, 1970).

[2] *Ibid.*, pp. 111–112.

[3] See Nelson Goodman, *Ways of Worldmaking* (Indianapolis: Hackett Publishing Co., 1978), Chs. 1, 7, particularly pp. 2, 110.

[4] See Joseph Margolis, 'Relativism, History, and Objectivity in the Human Studies,' *Journal for the Theory of Social Behavior* **XIV** (1984), 'Scientific Realism as a Transcendental Issue,' *Manuscrito* **VII** (1984).

[5] See for example 'Postscript – 1969' to *The Structure of Scientific Revolutions*; T. S. Kuhn, 'Reflections on My Critics,' in Imre Lakatos and Alan Musgrave (eds.), *Criticism and the Growth of Knowledge* (Cambridge: Cambridge University Press, 1970), and T. S. Kuhn, 'Second Thoughts on Paradigms,' in Frederick Suppe (ed.), *The Structure of Scientific Theories* (Urbana: University of Illinois Press, 1970).

[6] *The Structure of Scientific Revolutions*, 'Postscript – 1969,' p. 206.

[7] *Ibid.*

[8] See Imre Lakatos, 'Falsification and the Methodology of Scientific Research Programmes,' *Philosophical Papers*, Vol. 1, eds. John Worrall and Gregory Currie (Cambridge: Cambridge University Press, 1978); originally published in Lakatos and Musgrave, *loc. cit.*

[9] Karl R. Popper, 'Normal Science and Its Dangers,' Lakatos and Musgrave, *loc. cit.*

[10] 'Postscript – 1969,' p. 179. Kuhn is surely influenced here by the work of such friendly commentators as Margaret Masterman, 'The Nature of a Paradigm,' in Lakatos and Musgrave, *loc. cit.*

[11] *Ibid.*, pp. 54–55.

[12] 'Postscript – 1969,' pp. 207–208.

[13] 'Postscript – 1969,' p. 178f. See also, Gary Gutting (ed.) *Paradigms and Revolutions* (Notre Dame: University of Notre Dame Press, 1980).

176 JOSEPH MARGOLIS

[14] Not meant here in Stephen Toulmin's technical sense; cf. 'Does the Distinction between Normal and Revolutionary Science Hold Water?' Lakatos and Musgrave, *loc. cit.*

[15] See, for instance, Kuhn, *The Structure of Scientific Revolutions*, pp. 102–103; Paul Feyerabend, *Against Method* (London: NLB, 1975), Ch. 17. See also, Dudley Shapere, 'The Structure of Scientific Revolutions,' *Philosophical Review* LXXIII (1964); 'Meaning and Scientific Change,' in R. G. Colodny (ed.), *Mind and Cosmos: Essays in Contemporary Science and Philosophy* (Pittsburgh: University of Pittsburgh Press, 1966).

[16] 'Reflections on My Critics,' p. 268.

[17] *Ibid.*, p. 266.

[18] This bears directly on the papers by J. W. N. Watkins, Stephen Toulmin, and Popper particularly, in Lakatos and Musgrave, *loc. cit.*; also, those by Shapere, cited above.

[19] Kuhn himself worries about the issue, for example in 'Postscript – 1969,' pp. 205–210. See also, Popper, *loc. cit.* For a sample of the pernicious consequences of this linkage, see most of the papers in Martin Hollis and Steven Lukes (ed.), *Rationality and Relativism* (Cambridge: MIT Press, 1982). On the issue of relativism, See Joseph Margolis, 'The Nature and Strategies of Relativism,' *Mind* XCII (1983).

[20] See Joseph Margolis, 'Pragmatism without Foundations,' *American Philosophical Quarterly* XXI (1984).

[21] This undoubtedly helps to explain the unusual interest in the recent work of Richard Rorty: see for instance *Philosophy and the Mirror of Nature* (Princeton: Princeton University Press, 1979); and *Consequences of Pragmatism* (Minneapolis: University of Minnesota Press, 1982).

[22] Rudolf Carnap, 'Psychology in Physical Language,' trans. George Schick, in A. J. Ayer (ed.), *Logical Positivism* (Glencoe: Free Press, 1959), pp. 166–167. See also, Joseph Margolis, 'Schlick and Carnap on the Problem of Psychology,' in Eugene T. Gadol (ed.), *Rationality and Science* (Vienna: Springer-Verlag, 1982).

[23] See Carl G. Hempel, 'Logical Positivism and the Social Sciences,' in Peter Achinstein and Stephen F. Barker (eds.), *The Legacy of Logical Positivism* (Baltimore: Johns Hopkins University Press, 1969).

[24] Paul Oppenheim and Hilary Putnam, 'Unity of Science as a Working Hypothesis,' in Herbert Feigl *et al.* (eds.), *Minnesota Studies in the Philosophy of Science*, Vol. II (Minneapolis: University of Minnesota Press, 1958). See also, Robert L. Causey, *Unity of Science* (Dordrecht: D. Reidel, 1977), Ch. 6.

[25] Oppenheim and Putnam, *loc. cit.*; Causey, *loc. cit.* See also, Mario Bunge, 'Emergence and the Mind,' *Neuroscience* XI (1977); and 'Levels and Reduction,' *American Journal of Physiology* CIII (1977).

[26] See Carl G. Hempel, *Aspects of Scientific Explanation* (New York: Free Press, 1965), Ch. 4; *Philosophy of Natural Science* (Englewood Cliffs, 1966), Ch. 5.

[27] Causey, *loc. cit.*

[28] Otto Neurath, 'Protocol Sentences,' trans. George Schick, in Ayer, *loc. cit.*

[29] See Joseph Margolis, 'Pragmatism without Foundations'; 'Skepticism, Foundationalism, and Pragmatism,' *American Philosophical Quarterly* XIV (1977).

[30] Karl R. Popper, 'The Aim of Science,' *Objective Knowledge* (Oxford: Clarendon Press, 1972). Cf. also, Rorty, *loc. cit.*

[31] See Joseph Margolis, *Persons and Minds* (Dordrecht: D. Reidel, 1980; *Culture and Cultural Entities* (Dordrecht: D. Reidel, 1984; *Philosophy of Psychology* (Englewood Cliffs: Prentice-Hall, 1984).

[32] Cf. Hempel, *Aspects of Scientific Explanation*, pp. 376–425; Wesley C. Salmon, 'Statistical Explanation,' in Wesley C. Salmon *et al.*, *Statistical Explanation and Statistical Relevance* (Pittsburgh; University of Pittsburgh Press, 1971); Hans Reichenbach, *Laws, Modalities, and Counterfactuals* (Berkeley: University of California Press, 1976). On biological laws, see Ernst Mayr, *The Growth of Biological Thought* (Cambridge: Harvard University Press, 1982), Ch. 2; on the prospect of laws in the linguistic and cultural domains, see Margolis, *Culture and Cultural Entities*.

[33] W. V. Quine, *Word and Object* (Cambridge: MIT Press, 1960); Donald Davidson, 'In Defense of Convention T,' in Hugues Leblanc (ed.), *Truth, Syntax and Modality* (Amsterdam: North Holland, 1973); Donald Davidson, 'Mental Events,' in L. Foster and J. Swanson (eds.), *Experience and Theory* (Amherst: University of Massachusetts Press, 1970). See also, Ian Hacking, *Why Does Language Matter to Philosophy?* (Cambridge: Cambridge University Press, 1975); Gareth Evans and John McDowell (eds.), *Truth and Meaning* (Oxford: Clarendon Press, 1976).

[34] Margolis, *Culture and Cultural Entities*.

[35] See for example Hilary Putnam, *Meaning and the Moral Sciences* (London: Routledge and Kegan Paul, 1978); *Reason, Truth and History* (Cambridge: Cambridge University Press, 1981); Nelson Goodman, *loc. cit.*; Michael Dummett, *Truth and Other Enigmas* (Cambridge: Harvard University Press, 1978); Paul Ricoeur, *Hermeneutics and the Human Sciences*, ed. and trans. John B. Thompson (Cambridge: Cambridge University Press, 1981); Joseph Margolis, 'Scientific Realism as a Transcendental Issue.'

[36] Karl B. Popper, *Realism and the Aim of Science (Postscript to the Logic of Scientific Discovery*, Vol. I), ed. W. W. Bartley, III (Totowa, N.J.: Rowman and Littlefield, 1983), Ch. 1, 2, 4.

[37] See Karl R. Popper *Conjectures and Refutations*, 2nd ed. (London: Routledge and Kegan Paul, 1965), Ch. 10 and addenda.

[38] See Margolis, 'Pragmatism without Foundations.'

[39] See Reichenbach, *loc. cit.*

[40] See Feyerabend, *loc. cit.*

[41] Noam Chomsky, *Rules and Representations* (New York: Columbia University Press, 1980), p. 6. Cf. Jerry A. Fodor, *The Modularity of Mind* (Cambridge: MIT Press, 1983).

[42] *Ibid.*

[43] *Ibid.*, p. 16. Cf. the rest of Ch. 1; and Donald Hockney, 'The Bifurcation of Scientific Theories and Indeterminacy of Translation,' *Philosophy of Science* **XLII** (1975).

[44] See Margolis, *Culture and Cultural Entities*, Chs. 6–7.

[45] Noam Chomsky, *Language and Responsibility*, trans. John Viertel (New York: Pantheon Books, 1977), pp. 152–153.

[46] Claude Lévi-Strauss, *The Savage Mind*, in trans. (Chicago: University of Chicago Press, 1966), p. 247.

[47] *Ibid.*, pp. 248, 249, 252. See also, Joseph Margolis, 'The Savage Mind Totalizes,' *Man and World* **XVII** (1984).

[48] Jean Piaget, 'Piaget's Theory,' in Bärbel Inhelder et al. (eds.), Piaget and His School (New York: Springer-Verlag, 1976), pp. 11–12; this is a modified version of an earlier article by Piaget, originally published in P. H. Mussen (ed.), Carmichael's Manual of Child Psychology, 3rd ed. Vol. 1 (New York: John Wiley and Sons, 1970). See also, Jean Piaget, Structuralism, trans. Chaninah Maschler (New York: Basic Books, 1970); Biology and Knowledge (Chicago: University of Chicago Press, 1971).

[49] Ibid., p. 22.

[50] L. S. Vygotsky, Thought and Language, trans. Eugenia Hanfmann and Gertrude Vakar (Cambridge: MIT Press, 1962).

[51] See Massimo Piattelli-Palmarini (ed.), Language and Learning: The Debate between Jean Piaget and Noam Chomsky (Cambridge: Harvard University Press, 1980).

[52] See Jacques Lacan, Speech and Language in Psychoanalysis, trans. Anthony Wilden (Baltimore; Johns Hopkins University Press, 1968), originally published as The Language of the Self; and Jürgen Habermas, Knowledge and Human Interests, trans. Jeremy J. Shapiro (Boston: Beacon Press, 1971).

[53] See Joseph Margolis, 'Reconciling Freud's Scientific Project and Psychoanalysis,' in Daniel Callahan and H. Tristram Engelhardt, Jr. (eds.), Morals, Science and Sociality (Vol. III, The Foundations of Ethics and Its Relationship to Science (Hastings-on-Hudson: The Hastings Center, 1978); 'Goethe and Psychoanalysis,' in Francis J. Zucker, Frederick Amrine, and Harvey Wheeler (eds.), Goethe and the Sciences: A Reappraisal (Dordrecht: D. Reidel, 1986).

[54] See for instance Rudolf A. Makkreel, Dilthey; Philosopher of the Human Sciences (Princeton: Princeton University Press, 1975); Michael Ermarth, Wilhelm Dilthey: The Critique of Historical Reason (Chicago: University of Chicago Press, 1978).

[55] Hans-Georg Gadamer, 'The Heritage of Hegel,' Reason in the Age of Science, trans. Frederick G. Lawrence (Cambridge: MIT Press, 1981), p. 61.

[56] Jürgen Habermas, 'What Is Universal Pragmatics?' Communication and the Evolution of Society, trans. Thomas McCarthy (Boston: Beacon Press, 1979). See Thomas McCarthy, 'Rationality and Relativism: Habermas's 'Overcoming' of Hermeneutics,' in John B. Thompson and David Held (eds.), Habermas: Critical Debates (Cambridge: MIT Press, 1982); and Joseph Margolis, 'Historicism, Universalism, and the Threat of Relativism,' unpublished.

[57] Regardless of one's final assessment of Gouldner's thesis. See Alvin W. Gouldner, The Two Marxisms (New York: Seabury Press, 1980).

[58] Theodor W. Adorno, Negative Dialectics, trans. E. G. Ashton (New York: Seabury Press, 1973); Michel Foucault, The Order of Things (in translation) (New York: Random House, 1970); Paul Ricoeur, Hermeneutics and the Human Sciences, ed. and trans. John B. Thompson (Cambridge: Cambridge University Press, 1981).

[59] Jacques Derrida, Of Grammatology, trans. Gayatri Chakravorty Spivak (Baltimore: Johns Hopkins University Press, 1976); Feyerabend, loc. cit.

PETER T. MANICAS

EXPLANATION AND QUANTIFICATION

INTRODUCTION

In 1929 Herbert Hoover assembled a distinguished group of social scientists 'to examine into the feasibility of a national survey of social trends ... to undertake the researches and make ... a complete impartial examination of the facts'.[1] Funded by the Rockefeller foundation with support from the Social Science Research Council and the Encyclopedia of the Social Sciences, four years of work by hundreds of inquirers resulted in 1600 pages of quantitative research, *Recent Social Trends in the United States*, more informally, 'the Ogburn Report', after its director of research, the Columbia University sociologist William F. Ogburn.

I use this document to date what from one point of view could be called the maturation of social science, the full-fledged incorporation of quantitative techniques into social science. From another point of view, of course, one might also say that it represented the firm incorporation in the United States, of government, private foundations, the university, and Vienna-inspired philosophy of social science. Or finally, one might say that it represented an accommodation – a *confused* accommodation if I am correct – of what had been two entirely independent tasks, 'social research' and social science.

In this too brief introduction, I want to give some historical background to my main claim, that quantitative social science has a proper role to play but that it pretends to play a role which it could not possibly play. I begin by agreeing completely with Lazarsfeld, Oberschall, Reiss and others that modern sociology has 'two roots', 'social research' and 'social theory', or what I would call, 'explanatory social science'.[2] The first has early origins in 'state-tistics' but was dramatically enlarged in the nineteenth century, especially by reform-minded charity organizations, royal commissions, and semi-academic organizations like the *Verein für Sozialpolitik*. It is hard to imagine a modern state without means to gather information about society and it is hard to imagine that most of this information not be quantitative. The origins of 'social theory', of course, are still earlier and this is hardly the

179

Barry Glassner and Jonathan D. Moreno (eds.),
The Qualitative-Quantitative Distinction in the Social Sciences. 179–205.
© 1989 *by Kluwer Academic Publishers.*

place to even suggest the enormous range of competing conceptions of what this has meant, from Aristotle, to Montesquieu, Herder, Comte, Mill, Hegel and to the later nineteenth century figures, e.g., Herbert Spencer, Gustav Schmoller, Engels, Weber, Simmel, Durkheim and Pareto. As today there remain disagreements between 'social theorists' on the question of an explanatory social science, between these nineteenth century giants, there were very large differences on what this meant. Except to notice that both Durkheim and Pareto had visions of a quantitative *and* explanatory social science and thus are central figures in what came to be quantitative social science, I do not in this essay try to develop either their visions or the opposing visions of, e.g., Engels or Weber.[3] On the other hand, I do take a stand on what I mean by explanation.

The late nineteenth century, if I am correct, was critical in the assimilation of these two independent tasks. The increasing need for 'social research' but especially for authoritatively sanctioned and 'objective' research gave 'social scientists' in the new universities in the U.S. new opportunities. But these opportunities could not have been exploited without a further convergence: the complete victory in the last decades of the century by what must be termed 'positivist' philosophy of natural science. As physical science came of age (as industrialized science), there emerged a group of natural scientists turned philosophers, men who could and did speak authoritatively regarding the nature of that physical science which, for the first time in history, had become a fundamental force in the everyday lives of ordinary people. While this group, which included Kirchoff, Ostwald, Mach, Boltzman, Pearson, Poincaré and Duhem, was by no means in complete agreement, they shared in attributing the success of science to its total separation from metaphysics, the signal feature of 'positive' philosophy.

For present purposes, it may be sufficient to appeal to Duhem's influential, *La Théorie Physique: Son Objet, Sa Structure* (1906).[4] Duhem saw that two answers had been given to the question, what is the aim of physical theory? Either 'a physical theory ... has for its object the explanation of a group of laws experimentally established' or, it is 'an abstract system whose aim is to summarize and classify logically a group of experimental laws without claiming to explain these laws'. But why is 'explanation' to be rejected? Duhem offers: 'To explain ... is to strip reality of the appearances covering it like a veil, in order to see bare reality itself.' Explanation in this sense makes science subordinate to metaphysics. He concluded: 'A physical theory is not an explanation. It is a system of mathematical propositions, deduced from a small number of mathematical principles which aim to represent as simply, as completely, and as exactly as possible a set of experimental laws' (p. 19). Explanation *is* description in just the sense that to explain some proposition is to do nothing more than to *deduce* it from higher order propositions.

How then did this help promote the identification of social research and social science? Social research had always employed 'descriptions' but especially 'statistically controlled observations': from counting (e.g., a census) to the identification of rates (e.g., unemployment, suicide), to the identification of correlations between quantified data (e.g., income and education, poverty and crime). Social research could be, of course, good or bad, 'objective' or 'biased'. This was not the issue. Rather, the question was, could even good social research be 'science'?

First, 'theory' was critical to science. Moreover, as everyone knew, the lawlike uniformities of theoretical physics, as of any science, were tested 'experimentally'. On the one hand, then, if a *scientific* theory (in contrast to a non-scientific theory) was as Duhem said it was (and one had only to look at a book in mathematical physics to see that it was), then if social science was to be science, it had to abandon the *metaphysical* theories of nineteenth century 'social theorists', not merely the pretentions of the grand theories of history, Evolutionary, Hegelian and Marxist, but as well, it had to abandon the notion that there were 'underlying' causes, 'essences', and other explanatory fictions. Pareto put the point neatly: 'Scientific laws are ... nothing more than experimental uniformities ... Modern scientists study the movements of the stars directly, and go no further than required for establishing uniformities of such movements.' Surely, a social scientist could do this? Nor were explanatory fictions part of 'theory'. One may ask, he continues, 'What is gravitation?' but 'celestial mechanics can dispense with a solution to it.' Indeed, 'it can be said that establishing a theory is something like passing a curve through a number of fixed points.'[5] The scientific sociologist had at least to do as well (sic) as economics had done.

What then of the experimental aspect of 'experimental uniformities'? There were two possibilities. First, granting that there are especially severe moral constraints on social experimentation, some things could be done. One might (as Dewey often suggested), 'experiment' with policy. Change it and see if what happens is what is predicted to happen.[6] Plainly, there were problems with such an idea: not only were 'controls' absent but as well, one needed to describe more than impressionistically the situation both before and after the introduction of the policy change. But perhaps some of the practical restraints on this might be reduced if one could sharpen one's descriptive tools and gather sufficient resources to carry out a sufficiently detailed study – the dream of the quantitative vision. A much discussed 'experiment' of the recent past is reported by Kershaw and Fair (1976–77) testing the hypothesis that income subsidies reduce incentives to work. A sample of about 1300 low-income families was subsidized at differing rates over a three-year period. The conclusion, hotly contested, was that subsidies do not reduce work incentives (Rossi and Lyall, 1976).

But a possibility which did not require manipulation of the environment was even more accessible. As the Committee on Basic Research in the Behavioral and Social Sciences pointed out, one can *simulate* experimental conditions by investigating 'the net contribution of each of several variables to a given outcome.'[7] Ackoff is even clearer: 'The development of methods of 'multivariate analysis' has removed the necessity of manipulation and the laboratory, and it permits scientists to go out into the world and tackle increasingly complex problems in their natural habitat.'[8]

It is fair to say, I believe, that such inquiry constitutes the very heart and soul of 'social research' now conceived as explanatory social science. Thus, the older methods of multivariate anaysis, e.g., trend studies, panel analysis and prediction studies are, it is held, powerfully supplemented by various regression techniques, path analyses, and computed assisted multivariate analytic techniques. All these require the ability to quantify variables. The upshot of these analysis are law-like statements which, as in the physical sciences are explanatory in just the sense that Poincaré and Mach had said that $D = 1/2at^2$ is explanatory.[9] Of course, in the social sciences, systematization of these propositions by means of an appropriate mathematical model is by no means child's play. Still, work by Blumen *et al.* (1955), Hägerstrand (1960), Goodman (1965), Duncan (1966), Bartholomew (1967), McGinnis (1968), Blau (1970), Ginsberg (1971), Hamblin *et al.* (1973 and 1978), Coleman (1973), Freeman and Hannan (1975), Doreian and Hummon (1976), Hamilton and Hamilton (1981), and others demonstrate (ostensibly?) maturing possibilities.[10]

These most sophisticated and ambitious efforts do not (yet?) represent the whole of quantitative social science or still less the whole of that work which is governed by a commitment to 'empirical methods'. Nor plainly, did a vision of a 'factual-statistical' representation of the social world ever go unchallenged. For example, Sorokin had no objection to statistics or to the use of mathematics, but he thoroughly savaged the Ogburn report remarking:

In the future some thoughtful investigator will probably write a very illuminating study about these 'quantitative obsessions' ... of the first third of the twentieth century, tell how such a belief became a vogue, how social investigators tried to 'measure' everything; how thousands of papers and research bulletins were filled with tables, figures and coefficients; and how thousands of persons never intended for scientific investigation found in measurement and computation a substitute for real thought.[11]

Adolph Berle remarked that 'the desire for objectivity has been carried too far' and Charles Beard offered that 'the results ... reflect the coming (sic) crisis in the empirical method to which American social science has long been in bondage.'

This paper has as its focus so-called 'hard science' approaches to social science. But indeed, I think that it can be shown that even those who reject the 'hard science' methodologies of quantitative workers are enthralled by the assumptions of 'empirical method' but in particular by notions of causality and of explanation which are widely shared – notions which, if I am correct, are wholly untenable.[12]

In what follows, then, I will argue that the key feature of 'quantitative social science' is an orthodox Humean constant conjunction conception of causation aligned with a notion of theory as a deductive system, a fatal combination. After sketching this, I argue that confusion over description and explanation is propelled by the failure to see what is doing the real job of explaining and by a pervasive confusion between the analysis of variation and causal analysis. By contrast, I hold to a more familiarly commonsensical notion of causality and explanation, that social scientific explanation, like all scientific explanation, requires showing how what happened came to happen. In turn I suggest that this requires the development of theory in a sense radically different from the sense employed in behavioral research.

The present paper does not challenge the problematic assumption that any phenomenon can be quantified, nor does it deal with the many problems raised in generating useful indices for concepts employed in social research. Nor to repeat, do I deny that these efforts are extremely valuable. The results of inquirers engaged in 'social research' are absolutely indispensable to any social science. The point, however, is that relations of quantified variables are not and cannot be explanatory. Rather, they are descriptions to be explained and thus, count as evidences of the theories and accounts which explain them.

THE ORTHODOX CONCEPTION OF CAUSATION AND EXPLANATION

It will not be necessary to attempt a full-fledged analysis of the received view of causality.[13] The received (Humean) view holds that a causal statement has the form,

If C, then E

where C and E designate singulars: events, conditions, etc. The antecedent C, from an existential point of view, is an event or condition which is prior to or simultaneous with its effect and the relation between C and E is a constant conjunction: Whenever C, then E. The relation is not necessary. Nor is it supposed that C has some active or productive power which brings about E. On empiricist views, powers are dispositions defined in terms of hypotheticals. To say that salt is watersoluble is to say no more than that if salt is put in water, it dissolves.[14]

It is but a short step then to the mathematical language of dependent and independent variables. Thus, the writer of a widely used text says, 'the independent variable in a research study is the antecedent; the dependent variable is the consequent'.[15] And from here, it is easy to generalize: $Y = f(x)$ and thus to introduce many variables in relation. Indeed, 'if science is largely the study of relations then it is largely the study of sets of ordered pairs' (p. 49).

The point of this is explanation. 'The only way to explain anything ... is to determine how that thing relates to other things.' 'Explanation of prejudice means to find how prejudice is related to other natural phenomena' where 'natural phenomena' means occurrences in the observable world (p. 62). 'Any phenomena, to be natural phenomena, must be observable or potentially measurable or manipulable.' 'If numerals can be assigned to objects according to rules – then we call a construct a variable' (p. 40). Thus, prejudice is a disposition which 'implies behavior of some kind' and the problem is find how 'variables' relate to 'variables.'

In his book, Kerlinger provides a detailed account of an actual study (Majoribanks, 1972) in which the dependent variable was 'mental development', measured by four tests. There were two independent variables, 'environmental press' and ethnic group membership. Each of these had a number of measures. Multiple regressions yielded Table I.

Quoting Kerlinger's conclusion:

Taking the values of [Table I] at face value, we can reach two or three conclusions. Both environment and ethnicity seem to have considerable 'influence' on verbal ability, especially when they 'work together' (34 percent). Their contributions alone, while not large, are appreciable (11 percent and 16 percent) ... A similar analysis can be applied to reasoning ability. We note especially that environment and ethnicity are not nearly as strongly related to reasoning ability as verbal ability. It is not hard to understand this rather important (*sic*) finding. The reason is left to the reader to deduce (p. 176).

TABLE I

Dependent variable	Independent variable	R^2
Verbal ability	Environment + Ethnicity	0.61
	Environment	0.50
	Ethnicity	0.45
Effect of Ethnicity Alone		0.11
Effect of Environment Alone		0.16
Reasoning Ability	Environment + Ethnicity	0.22
	Environment	0.16
	Ethnicity	0.08
Effect of Ethnicity Alone		0.06
Effect of Environment Alone		0.14

Kerlinger is conscientious in warning his readers that with two or more independent variables, 'analysis and interpretation become much more complex, difficult and even elusive' (p. 176). Worse, like all methods, this method yields only estimates of the values of the R^2's. But finally and 'perhaps above all, researchers will be extremely cautious about making causal statements.'

Even though we used expressions like 'accounted for' and 'effects,' causal implications, while perhaps inescapable because of language connotations, were not intended ... When we talk about the influence of ethnicity on verbal ability, for example, we certainly intend the meaning that the ethnic groups to which a child belongs influences his verbal ability – for obvious reasons. The more accurate research statement is that there are differences in verbal ability between say, Anglo-Saxon Canadians and French Canadians. But this is a functional difference in ability in the English language. We do not mean that being Anglo-Saxon, in and of itself, somehow 'causes' better verbal ability in general than being French Canadian. The safest way to reason is probably the conditional statement emphasized through this book: If *p*, then *q*, with a relative absence of causal implication (p. 177).

The foregoing suggests a number of important points. It shows, first, that
methodologists in the social sciences are aware that a host of ordinary
language expressions have, as Kerlinger says, 'causal implications', that
indeed, 'influences' (notice the scare quotes in the foregoing), 'effects', 'is
due to', 'accounted for', and others ordinarily connote causal efficacy in
exactly the realist sense that causes *bring about their* effects.

This is the second point. On the standard (Humean view), Kerlinger should
not be uncomfortable since on this view, 'if *p*, then *q*' is sufficient for
imputing causality. And if so, then as the Marjoribanks study shows, since
'being Anglo-Saxon' is regularly associated with superior verbal ability, we
have the *causal* expression, 'If some one is Anglo-Saxon, then there is a
probability K that this person will have verbal ability superior to...' As Paul
Lazarsfeld, an eminent methodologist, long ago pointed out:

> If we have a relationship between '*x*' and '*y*' and if for any antecedent test factor the
> partial relationships between '*x*' and '*y*' do not disappear, then the original relation-
> ship should be called a causal one. It makes no difference here whether the necessary
> operations are actually carried through or made possible by general reasoning.[16]

Lazarsfeld, a theorist of social science, is here a consistent Humean. A matter
of fact connection warrants the imputation of causality for that is all that
causality can mean. Kerlinger, a pedagogue of social scientists, is uneasy.
While for him *in the ordinary sense*, 'the ethnic group to which a child
belongs influences his verbal ability – for obvious reasons', the 'more
accurate' research statement is that there is a measureable difference in verbal
ability between the two groups.

Finally, on the standard empiricist account of explanation, law-like
statements can function in predictive contexts. The explanation or prediction
takes the form of modus ponens: If *p*, then *q* (covering law), *p*, therefore *q*
(event to be explained or predicted). To be sure, as Kerlinger writes, 'such
explanations are necessarily only partial and incomplete' and equivalently,
predictions are not certain. This is because there are many variables and their
relations will be complex. There are 'influences' on verbal ability other than
ethnicity and no one pretends that all of them have been identified in their
exact relation to the dependent variable. Nevertheless, this is not dishearten-
ing since it sets the agenda for further research. As a psychologist faced with
the same problem recently pointed out:

> The only way psychologists will ever come to understand complex psychological
> causation is to analyse variables, one by one, sub-set by sub-set, until whole systems
> of variables are understood.[17]

I want to show that this research program is futile, but first we need to sketch a plausible alternative.

REALIST CAUSALITY AND EXPLANATION

We need to rid ourselves fully of the idea that causes are only 'events', efficient causes, sufficient conditions, and on the other hand, we must take seriously the full implications of the idea that 'effects' could ever be the outcome of *a* cause. Instead, then, of thinking of causes as about the relations of events or classes of events – regularity determinism – we should think of them as properties of structures that exist and operate in the world.[18] Everything then is the outcome of multiple causes operating in the world, from the most stable empirical property, the yellowness of gold, to an 'event' like combustion – or for that matter, the Great Depression. But, as this is crucial, causes never operate in complete closure.

For example, current chemical theory details the causal properties of elements and compounds and thus yields laws about what they have the power to do. The formula, '$2H_2 + O_2 = 2H_2O$' is a causal law in the sense that if the theory is true, $2H_2$ and O_2 *must* produce water. By virtue of what they are, hydrogen and oxygen gas have causal properties among which (in appropriate conditions) is the outcome, H_2O. I say 'must' here even though, of course, it may be that in combination, water does not result. But it is easy enough to see why. The 'law' assumes theoretical closure, an abstraction from the concrete world, never realized. Put roughly, we can say that the causal properties of structures defined by theory never operate in total isolation from other potentially non-constant, effective structures having causal properties. Indeed, what we called above 'appropriate conditions' are but *other* causes at work in the world.

As suggested, causal laws are invariances – the force of the 'must' in my example. But empirical regularities – even when they are universal generalizations – are generalizations, not 'causal laws'.[19] 'Gold is yellow' and '$D = 16t^2$', needs here to be *contrasted with* 'salt is watersoluble' and '$F = g(mm/2)$', which are both causal expressions. To be sure, gold has causal properties which figure in color phenomena; it emits photons of such and such a frequency. But it is yellow by virtue of this *and* the less well-understand mechanisms of human perception. If something is gold and our theory is correct, it will have just those optical properties (and not some other). But it may appear other than yellow (both to us and to other sorts of percipients). The analysis of 'D = 16t^2' is even more straightforward. Even under closure, it expresses *no* causal relation. Rather, under closure, it is true by virtue of gravitation, a causal property of all bodies in space-time.

This is also why we experience patterns, not empirical necessities. The moon is full every twenty-eight days (more or less), salt usually dissolves in water, inflation often attends full employment, some beer drinkers get nasty after two beers. The last two patterns are asserted with less confidence than the first two. It is clear enough why. The examples differ in the degree of closure of the pertinent 'system'. That is, there are enormous differences in what is involved in *producing* the respective outcomes. Consider the last. The human organism, like everything else in the world, is a complex of orderly complex systems, but the 'outputs' which interest us – behavior – depend upon some extremely complex and presently ill-understood causal transactions, beginning, of course, with those biochemical processes which make us organisms and culminating in those psychological and social processes which make us human. Alcohol will have its chemical effects on the cells and these are well understood. But chemistry cannot tell us why some drunk gets belligerent on some occasions and not on others, nor why some other drunk gets affectionate, usually or often.

Because the things of the world are complexes of structures with causal properties, there is stability in the world (and generalizations do often hold); but because they never operate in isolation, there is precariousness. Indeed, because causes have outcomes which in transaction have outcomes and so on endlessly, there is genuine novelty in the world. And for the same reason, the future is 'open', undetermined in the sense that while not anything can happen, what will happen is, in principle, unpredictable.[20] This also allows us to make sense of experiment and to recast the problem of explanation.

When we can, we experiment. We isolate as far as possible a theoretized mechanism to see if in near closure, it does what theory postulates it *should* do. The experiment is a 'controlled' situation in that it yields what *would* not have happened (or been observed) without the experiment. Experiment is not 'twisting the tail' to see what *might* happen, nor do we experiment because we are interested in getting more reliable, more precise and more probable empirical regularities. Rather, we experiment because we believe that causes operate even in the absence of experimental closure.

Kerlinger was quoted, earlier, to the effect that 'the only way to explain anything ... is to determine how that thing relates to other things.' This statement is unexceptional. The problem is, what does 'determine how that thing relates to other things' mean?

It will hardly be possible to deny that one *can* mean by this showing that some 'variable' is always (or usually) related to some other variable, for example, that some egg is greenish-blue because it is a robin's egg and all robin's eggs are greenish blue or, similarly, that the moon is full today because it is full every 28 days. Generalizations do allow us to order experience, a consideration of no small importance. When they are universal, they even serve as a response to wonder, even if, as no one denies, they leave utterly unexplained, how it is, for example, that *any* robin's egg is greenish blue.[21] But when the generalization is short of being a universal, we lose hold of the individual case. When we believe that there is no causal mechanism linking the 'variables', it is hard to see what has been accomplished. It is no explanation to be told that something is G because it is F and there is high correlation (merely) between F and G. On the other hand, to be told that Sam got a mild flu because some percent of those who are exposed to the virus contract flu (whether this be a low *or* high percent) is helpful (to the extent it is) *only* because we know that the virus played a causal role. But as Hempel has himself insisted, 'a condition that is nomically necessary for the occurrence of an event does not, in general explain it'.[22] If it did, we could explain a disasterous fire but pointing out that oxygen was present.

Indeed, as Salmon has argued, the problem arises even as regards explanations of the D-N form.[23] He was thus led to reconsider the assumption that explanation (or prediction) is an *argument* that renders the explanandum either a necessary consequence of premises (in deductive nomological explanation/prediction) or as probable consequence (as in so-called inductive-statistical explanation). Indeed, it is easy to construct counter-examples in which 'the 'explanatory' argument is not needed to make us see that the explanandum event was to be expected.' Thus:

John Jones avoided becoming pregnant during the past year, for he had taken his wife's birth control pills regularly and every man who regularly takes birth control pills avoids pregnancy.

Salmon argued that to overcome this sort of counter-example (and the others already adduced), we need to radically revise the Hempelian schema. Since the problem turns on the *factual irrelevance* of explanans to the explanandum – a feature not disallowed by viewing explanation as an argument or inference – we need to reject the idea that an explanation is an inference. He offers instead of D–N and I–S (inductive statistical), what he calls S–R (statistical relevance): 'An explanation is an assembly of facts statistically relevant to the explanandum, regardless of the degree of probability that results' (p. 11). For Salmon, an explanation is a set of probability statements (including as a limiting case, probabilities of 0 or 1), along with a statement specifying the partition to which the explanation event belongs. For example, to explain why Sam contracted paresis, we offer that the prior probability of anyone contracting paresis is very low, but there is a statistically relevant partition of people who have untreated latent syphilis and those who do not. Since the probability that a person with untreated latent syphilis is low, but considerably higher than the prior probability of people in general, it will be an explanation of Sam's problem to point out that he has untreated latent syphilis.

While we can fully support Salmon's rejection of the idea that explanation is a relation of premise to conclusion, we must note also that his version (significantly, now abandoned by him), preserves the symmetry of explanation and prediction. It is easy to say why. As with the standard model, explanation need not determine how it came to be that *Sam* got paresis (or the less disastrous flu!). It is sufficient that 'we know which [outcomes] to bet on, which to bet against, and what odds. We know precisely what degree of expectation is rational' (p. 78).

One might say, generously, that this is an explanation under conditions of limited knowledge, one might equally say, I believe, that while it guides 'expectation', Salmon's model offers no explanation at all. We still do not know why (how it was) that *Sam* got paresis. Sensitive to this, perhaps, Salmon revised his view and did so in an interesting and pertinent way. He subsequently argued that we 'have a bona fide explanation of an event if we have a complete set of statistically relevant factors, the pertinent probability values, *and* causal explanations of the relevant relations.'[24] Aronson makes the decisive point:

Suppose we wish to explain why GI Joe died of leukemia. According to Salmon, the explanation is in terms of S-R plus a causal account, namely GI Joe was in the vicinity of an atom bomb blast and the subsequent gamma radiation caused his leukemia *and* the odds of his getting leukemia under these conditions were 0.01. But what does such a probability add to the explanation that the causal account has not already covered?[25]

Indeed! While I do not here attempt a full-blown analysis of causal explanation, the rough idea seems clear enough. We need to show how in any particular case a particular causal configuration occurred that had just the achieved result. This requires resolution of the event into its components and a theoretical redescription of them – causal analysis. As part of this, it requires retrodiction to possible causes and the elimination of alternatives that might have figured in the particular configuration. Explanation thus requires knowledge of the causal properties of the configured structures *and* an account *through time* of the particular and changing configuration. Because our knowledge is limited, we will not have the full story. Whether it is a good or bad explanation will be a straightforward function of how close we can come to saying that we understand the causes and how they came together to produce the event.

On the regularity views of explanation, the world is a determined concatenation of contingent events; for the realist, it is a contingent concatenation of real structures. It is for this reason, finally, that we are often in a position to explain some occurrence when we could not have predicted it. That is, explanation and prediction are *not* symmetrical.

But it still may be said: 'Of course, unemployment, belligerence, differences in verbal ability, etc. are caused and, of course, theory is required to identify causes'. It may then be insisted that quantitative researchers armed with multiple regression, factor and path analysis are interested in 'theory' and are aiming at finding causes in the ordinary sense.

Surely, there are plenty of statements by social scientists to evidence this. Indeed, the terms about which self-conscious methodologists are so careful, expressions like 'influences', 'is due to', 'affects', 'accounts for the variance of' are symptomatic. But it does not follow that researchers are doing what they believe they are doing. It remains for me to show that they are not.

THE ROLE OF THEORY

Kerlinger says,

A *theory* ... is a set of interrelated constructs (variables), definitions, and propositions that presents a systematic view of phenomena by specifying relations among variables, with the purpose of explaining natural phenomena (p. 64).

As far as I know, no one has ever given a very clear example of a real social scientific theory spelled out so as to fit Kerlinger's definition. Everybody presumably has the general idea and that is sufficient.[26] In his book, Kerlinger describes a small theory (of his own contrivance) in which strong, weak or no relations are observed between four variables: 'intelligence', 'school achievement', 'motivation', and 'social class'. This small theory seems close to the one Marjoribanks employed in his study. It will suffice in any case. Presumably, a researcher recognizes a pattern and generates some hypotheses regarding the relations involved. She then seeks to specify quantitatively the variables involved and to discover, through analytic techniques, their precise relations. Theory will serve to 'interpret' the results.

Consider then the relation between 'intelligence' and 'school achievement'. We may guess that the correlation here will be quite high and positive. But is this merely a correlation? Consider the pair:

(1)   If a person scores well on standard intelligence tests, then there is a probability K that he or she will be a high achiever in school.

(2)   If salt is put in water, then there is a probability K that it will dissolve.

These are *formally* identical. As regards (2) there surely is a causal relation. Indeed, (2) is an implication of (3): Salt is water-soluble, where 'water-soluble' refers to a causal power of salt. In this case, moreover, we have theoretical knowledge of that causal power, of why it is that salt tends to dissolve in water.

Can similar things be said about (1)? Plainly, there is some sort of real (non-spurious) relation between school achievement and intelligence. It is not like (say) a high correlation between the price of eggs in China and sales of the *Wall Street Journal*. Is there anything corresponding to 'water-soluble'?

Now I do not for a moment doubt that there is and that, therefore, one can, *versus* anti-naturalism in the philosophy of social science, pursue the idea that social science searches for causes. But it will be a fundamental point of the following that the behavioral researcher, following the empiricist methodology set out by Kerlinger (and countless others!) forbids one to do exactly what is necessary in order to advance our knowledge about what it is that does correspond to 'water-soluble'.

The point involves a tangle which takes some unravelling. In a nutshell, Kerlinger, the pedagogue of social science research, grasps that understanding is what science is all about, but at the same time, his empiricist commitments to Humean causality and to theory as an 'interpreted' axiom system leads him to misconstrue 'understanding' and to identify it with prediction and control.

In the preceding example, 'water-soluble' refers to a causal power of salt, made intelligible by means of a grasp of theorized natural structures and their dispositional properties, to NaCl, to H_2O and their nomic relations. A moment's reflection will suggest that in (1) what corresponds to 'water-soluble' is *a social mechanism which depends upon a set of structured practices*, including e.g., the system of intelligence testing, the measures it employs, the nature of the tests and quizzes which are used to grade students for 'achievement', etc. Just as a theory of the chemical properties of things gives us an understanding of the pattern 'salt tends to dissolve in water', we need a theory about schools, the structuring of their practices, and the relation of these to causally relevant other structures, the class structure, the state, etc.

The 'small theory' represented in Kerlinger's picture is not wrong so much as it is seriously misleading. It does this exactly because we take for granted exactly what is needed to do the job of explaining. We 'know', however unreliably, something about the practices of schools. This is, in fact, 'the general reasoning' referred to by Lazarsfeld in the text previously quoted. Such 'general reasoning' does indeed make it plausible that sometimes relations between 'variables' are causal. *It is the real theory in the background of such analyses*, a real, but unstated theory, which fools us into believing that the theory which is up front, the 'partial relationships' between dependent and independent variables are in any way explanatory.

Behavioral researchers are misled in part because they think of their equations as being on a par with the mathematical representations familiar to physical theories, or least to the familiar construction of them in textbooks and standard accounts in the philosophy of science. Indeed, it is hard to say how much damage has been done by the formalist assumption that everything that a theory is can be put down as a string of sentences.[27] But one must consider what in physics or chemistry these representations represent, what they do and how they function in these sciences.

It is easily shown that theory in the physical sciences depends crucially on non-propositional ideas, on non-deductive relations and especially on intentional meanings and meaning relations. My examples, from molecular chemistry, were especially useful since in that science, the 'ideal cognitive objects' which are at the core of the theory, atoms and molecules, can be represented with iconic models. The sentential modelling of the mathematical theory, then, will derive from the iconic model which is taken to be a representation of some thing or process. The mathematics may outrun the representation. As in quantum field theory, all that remains is 'metaphorical meaning in which only the mathematical properties of the use of the concept in the context of the model (e.g., as a correlated vector pair, one circulating and the other polar) survive together with the dispositional sense of the concept.' While the disposition of salt to dissolve in water is grounded in the mechanisms imputed by molecular chemistry, further grounding of the dispositions of the subatomic particles 'has been left behind'.

What is easily forgotten is that training to be scientist involves learning all that which is unspoken, but without which science is impossible. It was to Kuhn's credit that this point was finally communicated to energetic methodologists, interested in emulating their colleagues in the physical sciences. But except for the idea that there are no 'theory neutral data', the implications of the rejection of the deductivist view of theory seems to have made no impact on behavioral researchers. One implication, already noticed, bears further emphasis. It is the suggestion that 'the surplus meaning' of typical theories offered by behavioral researchers is merely 'commonsense'. That is, if the meanings of the key concepts, symbols and principles are not part of the explicit representation, in the physical sciences, at least, we can say that the meanings learned are part of a *well established theory into which scientists are socialized*. In behavioral research, by contrast, these meanings are *the unarticulated meanings of uncritical common sense*. The point is central.

All of us, of course, are socialized into *society*. It would be amazing if commonsensical 'knowledge' did not play a vital role in providing the background meaning for 'hypotheses' in social science. It is, indeed, *unavoidable*. What is avoidable, however, is the *uncritical appropriation* of this stock of ideas, something which cannot be done if, as is the case, it is not even acknowledged that 'general reasoning' is playing this *explanatory* role. Theorizing the mechanism of water-solubility was the product of conscious theory, of building up conceptions of atomic structures, valency and the rest. But where is the conscious theory construction – and test – of the social mechanisms of school differential achievement? Notice that I am not saying that behavioral researchers have no such theory. Rather, I am saying that their theory is the taken-for-granted, 'general reasoning,' and that it is unarticulated and held uncritically. Doesn't everyone know that women do poorly in math, that Catholics are anti-intellectual, that motivation is the critical fact in success! And as noted, because in the background there is an altogether familiar theory, the formulae do seem to explain. 'The ethnic group to which a child belongs influences his verbal ability – for obvious reasons'.

But there is a further confusion. Kerlinger wrote that 'science is an enterprise exclusively concerned with knowledge and understanding of natural phenomena. Scientists want to understand things' (p. 3). If Kerlinger is talking about the ideal of what had best be called 'theoretical science' then this formulation is exactly correct. But it is striking that 'social research' should be so characterized or indeed, that behavioral science aimed at prediction and control should be thought of as aimed at 'understanding'! He would argue, perhaps, that this is a cavil, that unless understanding is being used in some special sense, in the way perhaps that *verstehen* sociologists use the term, understanding, prediction, and control are *inherently* related. Indeed, on the standard covering law model of explanation, explanation and prediction are logically indiscriminable.

I have argued that this model must be rejected; but more can be said about prediction and control. Advances in our understanding of the properties which lie behind the manifest phenomena of the world have given us greatly increased abilities to generate technologies. Examples abound, from the biochemical knowledge of growth which has yielded improved fertilizers to the physical knowledge of the atom which has yielded weapons of horrifying destruction. And in this sense, we can increasingly make 'nature' subject to our wills. Understanding is thus connected to 'control' if control is the capacity to put to our use natural processes; but control in this sense is not prediction.

On the other hand, we can predict very well without having the slightest understanding, and conversely, we can understand very well and be utterly unable to predict. We can predict when we have a reliable regularity, one that exists independently of our intervention or because of it. But in neither case need we understand why the pattern occurs, what are its causes, or in the case where we bring about some effect through some act, why our action has the predictable effect it has. Humans could predict rain by seeing that, generally, ominous clouds of a characteristic sort preceded it. They did not need an understanding of temperature and dewpoint, the physics of condensation, etc.

Perhaps this is obvious enough. But if it is, then why does Kerlinger assert, immediately after saying that 'science is ... exclusively concerned with knowledge and understanding', that scientists 'want to be able to say: If we do such-and-such, then so-and-so will happen' (p. 17). *Understanding has suddenly collapsed into instrumental control.*

It is clear why. If 'theory' is nothing more than 'relations of variables' and if variables 'must be observable or potentially measureable or manipulable', then *necessarily* understanding becomes synonymous with 'prediction' or with instrumental control. For purposes of methodological incantations, at least, there is no surplus meaning in theory and theory does *not* offer a representation of the structures of reality, structures and processes which when cognized, would give an understanding but not, in themselves, the ability to predict or control.

As in physical science, there is a theoretical analogue in current social science. To recur to our example, if one is interested in understanding differential school achievement, then one is best advised to consider the writings of Paul Goodman, John Holt, Jonathan Kozol, Charles Silberman, Stephen Steinberg and Samuel Bowles and Herbert Gintis. Their 'theories' are not 'sets of interrelated variables' and they do not look much like the mathematical models of textbook theories. Nevertheless, these writers are interested in explaining the patterns which the practices of school generate and they see that to do this, one must get an understanding of the connected social mechanisms which underly these.

Indeed, there is something to the suspicion, widely expressed, that quantitative methodologists in the social sciences, enamoured of the success of their colleagues in the physical sciences, have appropriated the appearance of natural science – the mathematics, the computed assisted methods of analysis, etc. – but have left behind most of the substance.

CAUSAL ANALYSIS AND THE ANALYSIS OF VARIATION

Readers may not yet be persuaded that the realist alternative is an alternative, that on the one hand, the works of the writers just cited seem hardly 'scientific' and on the other, that theory in behavioral research is aimed at discovering the causes of phenomena and that quantitative methods remain the best means to that end. I want now to argue that if the search for causes is the aim of the standard research program, that program is futile. The reason for this is straightforward: Causes are not additive and all the quantitative methods in current use *must* assume that they are.[28]

We can recur to the earlier example, Marjoribank's study. The basic question was: 'How do environmental force and ethnicity affect mental development' (Kerlinger, p. 173). As seems clear, mental development will be a causal product of a complex epigenetic process which begins with conception and ends (finally) with the death of the organism. A particular genome, itself the product of the conjunction of haploid sex cells is, in embryogenesis, the locus of continuous transactions both in itself and in relation to its 'environment'. It finally emerges and is then in continuous transaction with a human environment. Through both of these conditions, as the biologist Paul Weiss says, 'the latitude for epigenetic vagaries of the component elements on all levels ... is immense'.

Marjoribanks aims to tackle a piece of this complicated problem and aims to do this by discovering how each of the independent variables separately affects mental development and how they affect it in concert (p. 173). I earlier summarized the results of the regressions, but we need to say a word on how they were derived. The equations are of the form,

$$Y = a + b_1X_1 + b_2X_2,$$

where 'Y' is the dependent variable (e.g., mental development), 'a' is a constant, irrelevant for present purposes, 'X_1' and 'X_2' are the independent variables ('environmental force' etc.), and 'b_1' and 'b_2' are regression coefficients. As is obvious, as multipliers of the independent variables, they weight them. Moreover, as is also obvious, the independent variables are taken to have additive features.

Determining these coefficients is not easy but the details need not trouble us here except to emphasize that 'regression equations give the best possible prediction given sets of data. No other equation will give us as good a prediction' (p. 170). That is, the test of the adequacy of any particular equation will be its predictive success given the set of data.

The upshot then will be the determination of R, the coefficient of multiple correlation. This is determined by correlating two sets of scores, the Y's calculated ('predicted') from the regression equations and the actual Y's as specified by the independent measure of Y. Given R, then, we can calculate R^2, interpreted as expressing the variance of the dependent variable accounted for by the regression combination of the independent variables.

To recur to the study (see Table I above), the correlation between verbal ability, a measure of mental development, and the combination of environment and ethnicity (as measured by Marjoribank's various measures) was 0.78 (R) and thus R^2 was 0.61. This means, accordingly, that '61% of the variance of verbal ability was accounted for by environment and ethnicity in combination'. But as well, separate regressions, between verbal ability and environment and between verbal ability and ethnicity yielded 0.50 and 0.45 respectively. By subtracting these from 0.61 we get 'the separate effects' of environment and ethnicity: e.g., 16% of the variance in verbal ability is 'accounted for' by environment alone.

This perhaps unduly detailed summary unnecessary for those familiar with the technique, has at least made clear that if this is causal analysis, causes must be additive. But of course, except for mechanical causation, vectors of forces, causes are not additive. In consequence, the foregoing 'analysis' is almost totally *meaningless*.

The near meaninglessness of this 'competent and imaginative study' is perhaps most convincingly demonstrated by comparing it to typical studies in biology which study the relation of heredity and environment.

If we take a genotype, e.g., seeds replicated by inbreeding or cloning (this minimizes genotypic individuality – a dominant feature of human genotypes) and place them in various controlled environments, it is possible to establish rough tables of correspondence between phenotype on the one hand, and genotype-environmental combinations on the other. The results, *never predictable in advance* give what is called 'the norm of reaction' of that genotype, its 'range of reaction' to environmental variations. Now, it is not possible to predict these norms because – and this is crucial – genetic and environmental factors are not additive. They are *causes in transaction*, not independent and interacting like vectors of force. That is, because genes cause different outcomes in different transactions and because the developmental process is mediated and transactional, the latitude for vagaries, as Weiss noted, is immense. Of course, it is not unlimited and thus one can arrive empirically at norms of reaction in such cases.

Multiple regression techniques are *not* meaningless *given* that such norms *have been experimentally established*. Across a range of environments in which independent variables have been specified and controlled, one could relate the variances in outcomes with the changes in the independent variables. One could produce meaningful R^2's. This would not, it must be emphasized, provide the *proportion* of causation between components since causation does not suddenly become additive. One would have, however, a satisfactory analysis of variations.

There is a parallel to Marjoribank's study, yet there is a gigantic difference. As regards this project, there is no experimental means to establish norms of reaction. In the first place, we are morally limited. We cannot clone fetuses and establish them in strictly controlled environments. But were this permissable, there would still be no way to specify all the relevant environmental variables exactly because these are not independent. The social world is real enough, but the mere fact that *necessarily* the social world is mediated by consciousness makes it impossible to say how controlled changes are related to what stays the same and how then the new condition is *experienced* by agents. Compare here a change in the amount of nitrogen in the soil as it bears on the development of a cloned seed of grain and a change in the physical environment of an infant in development. Clearly, causes are profoundly mediated in development and even the language of transaction – instead of interaction – is a radical oversimplication of the causal reality.

Of course, social scientists, including here psychologists, have always known this and we may assume that this is *a* reason that prediction becomes the sole test of the adequacy of the measures and equations. We simply rework the specifications and relations, from 'predicted' to actual, until we get a good fit. But the justification in terms of 'good prediction' is profoundly reinforced by Humean assumptions about causality and by technocratic assumptions about 'explanation'.

On the covering law model, explanation and prediction are symmetrical and on this conception of explanation, it is necessary only that there is a constant relation between the independent and the dependent variables. Given the extraordinary limits of experimentation in the social sciences, it is no wonder, then, that regression techniques, path analysis, etc. are so attractive. Given all these assumptions, one can be a real scientist without having a theory and without ever doing a real experiment. All we need is data – plenty of data!

It has not been a point of the present essay to argue that quantified data or statistical methods are irrelevant, either to science or to policy-making. Good social science will use all the good data it can get its hands on. It will use it as a way of describing more accurately what needs to be better understood, and it will use it to test theories about the causal mechanisms which are represented by theory.

It may be useful here to conclude with a brief sketch of a recent example of social science which seems to me to exhibit clearly what I have been arguing for. In their *Too Many Women? The Sex Ratio Question*, M. Guttentag and P. F. Secord offer the powerful hypothesis that 'the number of partners potentially available to men or to women has profound effects on sexual behaviors and sexual mores, on patterns of marriage and divorce, childrearing conditions and practices, family stability, and certain structural aspects of society itself'.[29] This provocative but vague generalization is made precise through the use of all sorts of quantified data, from demographic data about sex ratios, to data about divorce rates, family size, income distribution, etc.

But the authors' main aim is to articulate a complicated and multi-dimensional theory which explains these patterns. In an ingenious and highly plausible way, the authors link sex ratios, a structural theory of power of the sort familiar to Marxist political economy, a theory of what they call 'dyadic power', social exchange theory and finally some hypotheses about how culturally specific factors affect the outcomes of the social mechanisms identified by the theories mentioned.

The argument, which is informally elaborated, is assessed by means of comparative inquiry into how these mechanisms can explain crucial relevant differences in various societies and periods, e.g., classical Athens vs. Sparta,

Medieval Europe and contemporary America, and various differences
between e.g., black and white Americans, Orthodox Jews and non-Jews. Last,
but not least, they show how, *in terms of their theory*, certain otherwise
anomalous paterns are to be expected, and further that if their theory is
correct (or nearly so), that certain heretofore *unnoticed* patterns should be
expected and can be evidenced by independent statistical inquiry. Guttentag
and Secord used a great deal of social research and did some of their own.
But their book is an achievement in social science, construed as the effort to
explain the patterns and events of the social world.

NOTES

[1] This is quoted from Dean R. Gerstein's Introduction to *Behavioral and Social
Science: Fifty Years of Discovery* (Washington, D.C.: National Academy Press,
1986). This volume, in commemoration of the fiftieth anniversary of the Ogburn
Report, displays the remarkable continuity in vision in American social science from
1929 until today. Which is not to say, of course, that contemporary writers are not
more sensitive to at least some of the problems posed by that vision.
[2] See here usefully, Anthony Oberschall (ed.), *The Establishment of Empirical
Sociology* (New York: Harper and Row, 1972). It should not be overlooked that the
very idea of 'science' was for a very long time a contested idea and that it did not
come to have its present meanings until late in the nineteenth century. This claim and
the discussion which follows derive from my *History and Philosophy of the Sciences*
(New York and Oxford: Basil Blackwell, 1987).
[3] Weber was always an opponent of 'positivism' in social science. But although it is
often supposed that this stemmed from his adoption of Dilthey's notion of *verstehen*,
such is emphatically not the case. On the contrary, Weber held that quantified
relations could not be the goal of a *concrete* science since in the first place, if they
were true they were likely to be wholly uninteresting, and second, because even if
they were of some interest, we would still need to explain them! He believed that
physics was an abstract science and could say something useful about *all* bodies. After
all, all bodies had masses and all moved. These relations could be represented in
mathematical expressions. But since for him (and rightly) social scientists were
interested in the particular, specific, historical configurations which constitute human
societies, an explanatory social science had to be concrete. His solution was the
construction of the concrete concept, the ideal type, in terms of which we attempt 'to
understand [a society's] development, which is concretely determined and its
necessarily concrete patterns'. For a full-fledged account of Weber and of the
persistent confusion over 'nomothetic' and 'idiographic' science, see my *History*,
pp. 127–140.
    It is striking that during his long association with the *Verein*, Weber and his
associates did a great deal of social research, but he never confused his efforts at data
gathering with his 'theoretical' efforts at explanation.

Lazarsfeld misses this when he remarks: '... it turns out that Max Weber, the patron saint of large-scale theory builders, was closely connected with quantitative investigations of contemporary social problems six times in his life. He and the scholars who worked under his direction reported at least 1000 pages of research which, in style and format, would not be easily distinguished from the pages of our contemporary sociological journals'. This is quoted from Lazarsfeld's Preface to Anthony Oberschall, *Empirical Research in Germany, 1848–1914* (New York: Basic Books, 1965), pp. v-vi.

[4] Pierre Duhem, *The Aim and Structure of Physical Theory*, trans. by Phillip Weiner (Princeton: Princeton University Press, 1954).

[5] Wilfredo Pareto, *Manual of Political Economy* (New York: Kelley, 1971), p. 31; *Mind and Society*, 4 Vols (New York: Harcourt, 1935), par. 99, 97, 92. See my *History*, pp. 148–52.

Pareto, of course, was a superb mathematical economist. His vision of sociology was simply to apply general equilibrium theory to society taken as a social system. He wrote: 'In order to thoroughly grasp the form of a society in its every detail it would be necessary first to know what all the various elements are, and then to know how they function – and that in quantitative terms ... The number of equations would have to be equal to the number of unknowns and would determine them exhaustively' *(Mind and Society*, par. 2062).

One should consider what this means. Society is a closed system of law-governed relations, fully deterministic in the same sense that the solar system is fully deterministic. I think that this is the implicit assumption of all mathematical models of the social system, but it would take another essay to show this. But see below.

[6] Dewey's ambiguous role in American social science stems, at least in part from his often optimistic (and naive) remarks about the absence of 'experimentation' in social policy. See my 'Pragmatic Philosophy of Science and the Charge of Scientism', *Transactions of the Charles S. Peirce Society*, 24 (Spring 1988).

[7] Robert M. Adams, Neil J. Smelser and Donald J. Treiman (eds.), *Behavioral and Social Science Research: A National Resource* (Washington, D.C.: National Academy Press, 1982), p. 30. The author(s) of the report are not named. The Committee was indeed distinguished. I discuss in some detail the meaningfulness of this idea of 'experiment'. Here we need only note the point of such inquiry.

[8] See Russell L. Ackoff's much used *The Design of Social Research* (Chicago: University of Chicago Press, 1973), p. 4. But many other examples could be cited. It is worth mentioning a third necessary convergence: the development of suitable methods. Francis Galton (1889) developed the index of correlation, 'r' which was brought to its present form by Karl Pearson, Galton's successor at the Galton Laboratory for National Eugenics and the British trumpet for Mach's positivist philosophy. For a superb account of the development of modern statistics, its relation to eugenics, and of the problematic assumptions and debates within evolving statistical theory (comprehended as the articulation of a suitable framework for the analysis of numerical data), see Donald A. MacKenzie, *Statistics in Britain 1865–1930* (Edinburgh: Edinburgh University Press, 1981). MacKenzie rightly remarks: 'Pearson's methodological doubts about Galton's approach could be seen as exemplifying, rather than contradicting, the positivist and phenomenalist criteria of valid knowledge of the *Grammar*. Using statistics, the biologists could (apparently)

measure without theorizing, summarize facts with going beyond them, describe without explaining' (p. 88–89).

[9] Mach argued, e.g., that 'knowing the value of the acceleration of gravity, and Galileo's laws of descent, we possess simple and compendious directions for reproducing in thought all possible motions of a falling body. A formula of this kind is a complete substitute for a full table' (*Popular Scientific Lectures*, 5th Edition, Lasalle: Open Court, 1945, p. xl.).

[10] These examples are drawn from the useful essay by Raymond Boudon, 'Mathematical and Statistical Thinking in the Social Sciences', in the remarkable *Advances in the Social Sciences, 1900–1980*, edited by Karl W. Deutsch, Andrei S. Markovits and John Platt (Lanham: University Press of America, 1986). Boudon notes that a last, but 'not least' field of 'utmost importance' is data analysis and that 'much of the research activity of mathematicians and statisticians active in the social sciences, as well of some of the social scientists interested in 'methodology,' is devoted to developing new methods and techniques for the analysis of data' (p. 211). As regards regression with one or more equations, factor analysis, log-linear models, interaction analysis, and so on, he notes that 'although these methods are widely used, their exact function is sometimes ambiguous'. He concludes that 'data analysis methods are of crucial importance in empirical research, even though their main function has to be considered as *descriptive* rather than *explanatory* (p. 212). Compare here the 1905–1913 debate between Pearson and Yule, in MacKenzie, chapter 7. Briefly, Pearson held that there was an analogy in the correlation of nominal variables ('intelligence', e.g.), to interval-level variations with a joint normal distribution. This last had clear predictive meaning; if one could build an effective model for nominal variables and non-normal distributions which had predictive value, what was the objection? For Yule, 'measuring association' 'meant simply trying to summarize the degree of dependence manifest in a given nominal data' (p. 167). See below.

[11] Cited from Neil J. Smelser, 'The Ogburn Vision Fifty Years Later', in *Behavioral and Social Science, Fifty Years of Discovery*, p. 27. Smelser can no longer share in the naive positivist dream of theory-neutral 'objective facts', but he remains very much committed to the positivist conception of theory and explanation. He cites R. K. Merton's influential essays, 'The Bearing of Sociological Theory on Empirical Research' and 'The Bearing of Empirical Research on Sociological Theory', both in Merton, *Social Theory and Social Structure*, revised and enlarged edition (New York: Free Press, 1957). Thus, 'empirical generalizations' differ from 'so-called scientific laws' in that the latter 'is a statement of invariance *derivable* from a theory'. See below.

[12] See Douglas A. Porpora, *The Concept of Social Structure* (Denver: Greenwood Press, 1987), a beautifully developed analysis and critique of Durkheimian influenced 'nomothetic empiricist' social science à la Merton, Blau, Mayhew, etc. More surprising, perhaps, is the extent to which these notions (and contemporary analytic techniques for identifying causes) infect a writer as different from the foregoing as Theda Skocpol. See my review of her much discussed *States and Social Revolutions*, in *History and Theory*, 10 (1981).

[13] The orthodox view is classically discussed by Bertrand Russell, 'On the Notion of Cause, With Application to the Free-Will Problem' and by Herbert Feigl, 'Notes on Causality', both in H. Feigl and M. Brodbeck, *Readings in the Philosophy of Science*

(Berkeley: University of California Press, 1951). More recently, see John Earman, 'Causation: A Matter of Life and Death', *Journal of Philosophy*, 73 (1976); Thomas Beauchamp and Alexander Rosenberg, *Hume and the Problem of Causation* (New York: Oxford University Press, 1981).

[14] Of course, the literature on dispositions is also vast and I am aware that Carnap's original treatment was amended several times. See classically, Rudolf Carnap, 'Testability and meaning' (1963), reprinted in part in Feigl and Brodbeck, *Readings in the Philosophy of Science*. Characteristically, the implications of these withdrawals have not impacted on methodologists in the social sciences.

[15] Fred N. Kerlinger, *Behavioral Research: A Conceptual Approach* (New York: Holt, Rinehart and Winston, 1979), p. 21.

[16] Paul F. Lazarsfeld, 'Interpretation of Statistical Relations as a Research Operation', in Lazarsfeld and Morris Ginsberg (eds.), *The Language of Social Research* (New York: Free Press, 1955), p. 125.

[17] Walter L. Stroud, Jr., 'Biographical Explanation is Low-Powered Science', *American Psychologist*, 39 (1984).

[18] I make no effort here to provide an analysis of causality adequate for a realist theory. My sketch owes to Rom Harré and Edward Madden, *Causal Powers: A Theory of Natural Necessity* (Totowa, N.J. Rowman and Littlefield, 1975); Roy Bhaskar, *A Realist Theory of Science*, 2nd Edition (Atlantic Highlands, N.J.: Humanities Press, 1978); Gerald L. Aronson, *A Realist Philosophy of Science* (New York: St. Martin's Press, 1984).

[19] There is perhaps no better test of the fatal flaw of empiricist philosophies of science than their inability to sustain a distinction between 'laws of nature' and 'mere generalizations'. Every major nineteenth century writer struggled with the problem, resolved in our day, presumably, by the idea that 'laws' can be deduced ('derived') from higher order propositions. Of nineteenth century treatments, perhaps surprisingly, Weber's account was the most satisfactory. See my *History and Philosophy of the Social Sciences*, pp. 130–133.

[20] Bhaskar points out that two typical empiricist research programs, 'interactionism' and 'reductionism' flow from the (wild!) supposition that the universe is a closed system. He writes: 'since in the first case there are at the limit no conditions extrinsic to a system a full causal account would seem to entail a complete state-description (or a complete history) of the world. Similarly, as in the second case there can be no conditions intrinsic to the thing a causal statement entails a complete reduction of things into their presumed atomistic components (or original conditions)' (*A Realist Theory*, p. 77). Compare Pareto, above.

[21] See Fred I. Dretske, 'Laws of Nature', *Philosophy of Science*, 44 (1977).

[22] Carl C. Hempel, *Aspects of Scientific Explanation* (New York: Free Press, 1965, p. 369).

[23] Wesley C. Salmon, *Statistical Explanation and Statistical Relevance* (Pittsburgh: University of Pittsburgh Press, 1971). Salmon agrees with Dretske that entailment is the wrong relation for explanation; but they draw very different conclusions from this.

[24] 'Why Ask, 'Why'', Presidential Address, *American Philosophical Association*, Vol. 51 (1978).

[25] Aronson, *A Realist Philosophy of Science*, p. 198.

[26] Examples of efforts to set out what a sociological theory is supposed to look like

204   PETER T. MANICAS

include: R. K. Merton, *Social Theory and Social Structure* (New York: Free Press, 1957), Part I; Llewelyn Gross, 'Theory Construction in Sociology: A Methodological Inquiry' and more ambitiously, Joseph H. Greenberg, 'An Axiomatization of the Phonological Aspect of Language', both in Gross (ed.), *Symposium on Sociological Theory* (New York: Harper and Row, 1959). Perhaps the best statement of the idealized conception of theory, standard in recent neopositivism is Richard Rudner's account in *Philosophy of the Social Sciences* (Englewood Cliffs, N.J.: Prentice Hall, 1966). For a sociologist's version of this, see Jack Gibbs, *Sociological Theory Construction* (Hinsdale, Ill.: Dryden, 1972).

[27] This criticism was first developed by Rom Harré in his ground-breaking and too little appreciated, *Principles of Scientific Thinking* (Chicago: University of Chicago Press, 1970). The paragraphs which follow owe to this book and to his more recent *Varieties of Realism* (New York and Oxford: Basil Blackwell, 1986).

[28] My discussion follows Richard C. Lewontin, 'The Analysis of Variance and the Analysis of Causes' and Jerry Hirsch, 'Behavior-Genetic Analysis and Its Biosocial Consequences', both in G. Dworkin (ed.), *The IQ Controversy* (New York: Pantheon, 1976).

[29] Marcia Guttentag and Paul F. Secord, *Too Many Women? The Sex Ratio Question* (Beverly Hills: Sage, 1983), p. 9.

REFERENCES

D. J. Bartholomew (1967), *Stochastic Models for Social Process* (New York: Wiley).
P. Blau (1970), 'A Formal Theory of Differentiation in Organization', *American Sociological Review*, 35.
I. Blumen, M. Kogan, and P. J. McCarthy (1955), 'The Industrial Mobility of Labor as a Probability Process', *Cornell Studies in International Relations* (Ithaca: Cornell).
J. S. Coleman (1973), *The Mathematics of Collective Action* (London: Heineman).
P. Doreian and N. P. Hummon (1976), *Modeling Sociological Processes* (Amsterdam: Elsevier).
O. D. Duncan (1966), 'Methodological Issues in the Analysis of Mobility Tables' in N. Smelser and S. M. Lipset (eds.), *Social Structures and Mobility in Economic Development* (Chicago: Aldine).
J. H. Freeman and M. T. Hannan (1975), 'Growth and Decline in Organizations', *American Sociological Review*, 40.
L. Goodman (1965), 'On the Statistical Analysis of Mobility Tables', *American Journal of Sociology*, 70.
T. Hagerstrand (1960) 'The Monte-Carlo Simulation of Diffusion' in W. L. Garrison and D. F. Marble (eds.) *Quantitative Geography* (Evanston: Northwestern University Press, 1960).
R. L. Hamblin *et al.* (1973), *A Mathematical Theory of Social Change* (New York: Wiley).
J. D. Hamilton and L. C. Hamilton (1981), 'Models of Social Contagion', *Journal of Mathematical Sociology*, 8.

D. Kershaw and J. Fair (1967–77), *The New Jersey Income-Maintenance Experiment*, Three Vols. (New York: Academic Press).

K. Majoribanks (1972) 'Ethnic and Environmental Influences on Mental Abilities', *American Journal of Sociology*, 78.

R. McGinnis (1968), 'A Stochastic Model of Social Mobility', *American Sociological Review*, 33.

P. H. Rossi and K. C. Lyall (1976), *Reforming Public Welfare: A Critique of the Negative Income Tax Experiment* (New York: Russell Sage Foundation).

LIST OF WORKS CITED

Ackoff, R. L. *The Design of Social Research*. University of Chicago Press, Chicago (1973).

Adams, R. M., Smelser, N. J. and Treiman, D. J. eds. *Behavioral and Social Science Research: A National Resource*. National Academy Press, Washington (1982).

Adorno, T. W. *Negative Dialectics*. Ashton, E.G. trans. Seabury Press, New York (1973).

Alker, H. From information processing research to the sciences of human communication. *Informatique et Sciences Humaines* **40–41**: 407–420, 1979.

Alker, H. Logic, dialectics, politics: Some recent controversies. In: Alker, H. ed. *Poznan Studies in the Philosophy of the Sciences and Humanities*. Rodopi, Amsterdam (1982).

Alker, H. Polimetrics: Its descriptive foundations. In: Greenstein F. and Polsby N. eds. *Handbook of Political Science*. Addison-Wesley, Reading (1975).

Alker, H. Research paradigms and mathematical politics. In: Wildenmann, R. ed. *Sozialwissenschaftliches Jahrbuch für Politik*. Günter Olzog Verlag, Munich (1976).

Alvarez, A. *The Savage God*. Bantam, New York (1970).

Anderson, E. A place on the corner. University of Chicago Press, Chicago (1978).

Armstrong, D. *The Political Anatomy of the Body*. Cambridge University Press, Cambridge (1983).

Aronson, G. L. *A Realist Philosophy of Science*. St. Martin's Press, New York (1984).

Baker, K. M. *Condorcet*. University of Chicago Press, Chicago (1975).

Baker, K. M. The early history of the term 'social science'. In: Annals of Science, 20–, 1964.

Bar-Hillel, Y. Indexical expressions. *Mind* **63**:359–79, 1954.

Bartholomew, D. J. *Stochastic Models for Social Process*. Wiley, New York (1967).

Barton, A. H. and Lazarsfeld, P. F. Qualitative measurement in the social sciences. In: Lerner, D. and Lasswell, H. D. *The Policy Sciences*. Stanford University Press, Stanford (1951).

Beauchamp, T. and Rosenberg, A. *Hume and the Problem of Causation*. Oxford University Press, New York (1981).

Becker, H. Whose side are we on. Presidential Address to the Society for the Study of Social Problems, 1966.

Becker, H. S. and Geer, B. Participant observation: The analysis of qualitative field data. In: Adams, R. M. and Preiss, J. J. eds. *Human Organization Research: Field Relations and Techniques*. Dorsey Press, Homewood, IL (1960).

Becker, H. Problems of inference and proof in participant observations. *American Sociology Review* **23**(6):652–660, 1958.

Bernstein, R. *The Restructuring of Social and Political Theory*. University of Pennsylvania Press, Philadelphia (1978).

207

Barry Glassner and Jonathan D. Moreno (eds.),
The Qualitative-Quantitative Distinction in the Social Sciences. 207–224.
© 1989 *by Kluwer Academic Publishers.*

Berry, J. W. On cross-cultural comparability. *International Journal of Psychology* 4:119–128, 1969.

Bhaskar, R. *The Possibility of Naturalism.* Harvester, Brighton (1979).

Bhaskar, R. *A Realist Theory of Science.* Harvester, Leeds (1975).

Bhaskar, R. *A Realist Theory of Science.* 2nd ed. Humanities Press, Atlantic Highlands, NJ (1978).

Biernoff, D. Psychiatric and anthropological interpretations of 'aberrant' behavior in an Aboriginal community. In: Reid, J. ed. *Body, Land and Spirit.* University of Queensland, St. Lucia (1982).

Blau, P. A formal theory of differentiation in organization. *American Sociological Review* 35 (1970).

Blumen, I., Kogan, M. and McCarthy, P. J. The industrial mobility of labor as a probability process. In: *Cornell Studies in International Relations.* Cornell, Ithaca (1955).

Blumer, H. *Symbolic Interactionism: Perspective and Method.* University of California Press, Berkeley (1969).

Boas, F. *Handbook of American Indian Languages.* (BAE-B40, part I). Smithsonian Institute, Washington, DC (1911).

Boas, F. Recent anthropology. *Science* 98: 311–14, 334–37, 1943.

Bogdan, R. and Taylor, S. J. *Introduction to Qualitative Research Methods.* John Wiley and Sons, New York (1975).

Bosk, C. *Forgive and Remember.* University of Chicago Press, Chicago (1979).

Boudon, R. Mathematical and statistical thinking in the social sciences. In: Deutsch, K. W., Markovits, A. S. and Platt, J. *Advances in the Social Sciences, 1900–1980.* University Press of America, Lanham (1986).

Bougle, C. *The French Conception of 'Culture Generale' and Its Influences upon Instruction.* Columbia University, New York (1938).

Bourdieu, P. *Outline of a Theory of Practice.* Cambridge University Press, Cambridge (1977).

Boyton, G. R. *Mathematical Thinking about Politics: An Introduction to Discrete Time Systems.* Longman, New York (1980).

Braithwaite, R. B. *Scientific Explanation: A Study of the Function Theory, Probability and Law Science.* Cambridge University Press, Cambridge (1953).

Braybrooke, D. *Philosophy of Social Science.* Prentice-Hall, Englewood, NJ, (1987).

Brodbeck, M. ed. *Readings in the Philosophy of the Social Sciences.* Macmillan, New York, (1958).

Brown, C. H. Psychological, semantic and structural aspects of American English kinship terms. *American Ethnologist* 1: 415–36, 1974.

Brown, G. Some thoughts on grounded theory. *Sociology* 7(1):1–16, 1973.

Brown, S. Intensive analysis in political research. *Political Methodology* 1:1–25, 1974.

Bunge, M. Emergence and the mind. *Neuroscience* XI, 1977.

Bunge, M. Levels and reductions. *American Journal of Physiology* CIII, 1977.

Burr, A. The ideologies of despair: A symbolic interpretation of punks' and skinheads' usage of barbiturates. *Social Science and Medicine* 19(9):929–38, 1984.

Cain, M. and Finch, J. Towards a rehabilitation of data. In: Abrams, P. *et al. Practice and Progress: British Sociology 1950–80.* Allen and Unwin, London (1981).

Camic, C., The matter of habit. *American Journal of Sociology* **91**: 1039–87, 1986.

Cantore, E. *Scientific Man.* ISH Publishers, New York, (1977).

Carnap, R. Psychology in physical language. Schick, G. trans. In: Ayer, A. J. ed. *Logical Positivism.* Free Press, Glencoe (1959).

Carnap, R. Testability and meaning. In: Feigl, H. and Brodbeck, M. eds. *Readings in the Philosophy of Science.* University of California Press, Berkeley (1951).

Causey, R. L. *Unity of Science.* D. Reidel, Dordrecht (1977).

Caws, P. Reform and revolution. In: Virginia, H. *et al.* eds. *Philosophy and Political Action.* Oxford University Press, New York (1972).

Chomsky, N. *Language and Responsibility.* Viertel, J. trans. Pantheon Books, New York (1977).

Chomsky, N. *Rules and Representations.* Columbia University Press, New York (1980).

Cicourel, A. *Method and Measurement in Sociology.* Free Press, New York (1964).

Coleman, J. S. *The Mathematics of Collective Action.* Heineman, London (1973).

Collingwood, R. *The Idea of History.* Oxford Press, Oxford (1946).

Condorcet, M. de, *Condorcet: Selected Writings.* Baker, K. M. ed. Bobbs-Merrill, Indianapolis, (1976).

Crick, M. *Explorations in Language and Meaning.* John Wiley and Sons, New York (1976).

Dahl, R. *A Preface to Democratic Theory.* University of Chicago Press, Chicago (1956).

Daniels, A. K. Military psychiatry: The emergence of a subspecialty. In: Friedson, E. and Lorber, J. eds. *Medical Men and Their Work.* Aldine-Atherton, Chicago (1972).

Davidson, D. A coherence theory of truth and knowledge. In: Henrich, D. ed. *Kant oder Heigel?* Klett-Cotta, Stuttgart (1983).

Davidson, D. In defense of convention T. In: Leblanc, H. ed. *Truth, Syntax and Modality.* North Holland, Amsterdam (1973).

Davidson, D. Mental events. In: Forster, L. and Swanson, J. eds. *Experience and Theory.* University of Massachusetts Press, Amherst (1970).

Davidson, D. On the very idea of a conceptual scheme. In: *Inquiries into Truth and Interpretation.* Oxford Press, Oxford (1984).

Denzin, N. *The Research Act in Sociology.* Butterfield, London (1970).

Derrida, J. *Of Grammatology.* Spivak, G. C. trans. Johns Hopkins University Press, Baltimore (1976).

Dietz, M. Trapping the prince: Machiavelli and the politics of deception. *American Political Science Review* **80**:777–799, 1986.

Dingwall, R. The ethnomethodological movement. In: Payne, G., Dingwall, R., Payne, J. and Carter, M. eds. *Sociology and Social Research.* Croom Helm, London (1981).

Dingwall, R. and Murray, T. Categorization in accident departments: 'Good' patients, 'bad' patients and children. *Sociology of Health and Illness* **5**(2):127–48, 1983.

Doreian, P. and Hummon, N. P. *Modeling Sociological Processes.* Elsevier, Amsterdam (1976).

Douglas, J. *Investigative Field Research.* Sage, Beverly Hills (1976).

Douglas, M. *Implicit Meanings: Essays in Anthropology*. Routledge and Kegan Paul, London (1975).

Douglas, M. *Natural Symbols: Explorations in Cosmology*. 2nd ed. Barrie and Jenkins, London (1973).

Douglas, M. Self-evidence. In: *Implicit Meaning*. Routledge and Kegan Paul, London (1975).

Dretske, F. I. Laws of nature. *Philosophy of Science* **44** (1977).

Duhem, P. *The Aim and Structure of Physical Theory*. Weiner, P. trans. Princeton University Press, Princeton (1954).

Dummett, M. *Truth and Other Enigmas*. Harvard University Press, Cambridge (1978).

Duncan, O. D. Methodological issues in the analysis of mobility tables. In: Smelser, N. and Lipset, S. M. eds. *Social Structures and Mobility in Economic Development*. Aldine, Chicago (1966).

Durkheim, E. *The Division of Labor in Society*. Free Press, New York (1893, 1933).

Durkheim, E. The dualism of human nature and its social conditions. In: Bellah, R. ed. *Emile Durkheim on Morality and Society*. University of Chicago Press, Chicago (1914, 1973).

Durkheim, E. *Incest: The Nature and Origin of the Taboo*. Lyle Stuart, New York (1897, 1963).

Durkheim, E. *Moral Education*. Free Press, Glencoe (1925, 1961).

Durkheim, E. *Professional Ethics and Civic Morals*. Greenwood Press, Westport (1950, 1983).

Durkheim, E. *The Rules of Sociological Method*. Free Press, New York (1924, 1974).

Durkheim, E. Review of Ferdinand Tönnies, Gemeinschaft und Gesellschaft. In: Traugott, M. ed. *Emile Durkheim on Institutional Analysis*. University of Chicago Press, Chicago (1978).

Durkheim, E. *Socialism and Saint-Simon*. Antioch Press, Yellow Springs (1928, 1958).

Durkheim, E. *Suicide*. Free Press, New York (1897, 1951).

Earman, J. Causation: A matter of life and death. *Journal of Philosophy* **76** (1976).

Edelman, M. *Political Language*. Academic Press, New York (1977).

Edelman, M. *The Symbolic Uses of Politics*, 2nd ed. University of Illinois Press, Urbana (1985).

Ellenberg, H. *The Discovery of the Unconscious*. Basic Books, New York (1970).

Ember, C. R. Cross-cultural cognitive studies. In: Siegel *et al.* eds. *Annual Review of Anthropology* **6**: 35–56. Annual Reviews, Inc, Palo Alto (1977).

Engels, F. *Herr Eugen Duhring's Revolution in Science*. Burns, E. trans. International Publishers, New York (1939).

Ermarth, M. *Wilhelm Dilthey: The Critique of Historical Reason*. University of Chicago Press, Chicago (1978).

Eulau, H. The maddening methods of Harold D. Lasswell: Some philosophical underpinnings. In: Rogow, A. ed. *Politics, Personality, and Social Science in the Twentieth Century: Essays in Honor of Harold D. Lasswell*. University of Chicago Press, Chicago (1969).

Evans, G. and McDowell, J. eds. *Truth and Meaning*. Clarendon Press, Oxford (1976).

Feigl, H. Notes on causality. In: Feigl, H. and Brodbeck, M. eds. *Readings in the Philosophy of Science*. University of California Press, Berkeley (1951).

Feleppa, R. *Convention, Translation, and Understanding*. State University of New York Press, Albany (1988).

Feleppa, R. Emics, etics, and social objectivity. *Current Anthropology* 27:243–255, 1986.

Feleppa, R. Physicalism, indeterminacy, and interpretive science. *Metaphilosophy* (forthcoming).

Feleppa, R. Translation as rule-governed behavior. *Philosophy of the Social Sciences* 12:1–31, 1982.

Fenno, R. Notes on method: Participant observation. Appendix to: *Home Style: House Members in Their Districts*. Little, Brown, Boston (1978).

Fenno, R. Observation, context, and sequence in the study of politics. *American Political Science Review* 80: 3–15, 1986.

Feyerabend, P. *Against Method*. NLB, London (1975).

Fielder, J. *Field Research*. Jossey-Bass, San Francisco (1978).

Filmer, P. *et al. New Directions in Sociological Theory*. Collier-Macmillan, London (1972).

Fisher, L. E. and Werner, O. Explaining explanation: Tension in American anthropology. *Journal of Anthropological Research* 34:194–218, 1978.

Fisher, S. Doctor-patient communication: A social and micro-political performance. *Sociology of Health and Illness* 6(1):1–29, 1984.

Flew, A. *Thinking about Social Thinking: The Philosophy of the Social Sciences*. Basil Blackwell, New York (1986).

Fodor, J. A. *The Modularity of Mind*. MIT Press, Cambridge (1983).

Fodor, J. A. *Representations: Philosophical Essays on the Foundations of Cognitive Science*. MIT Press, Cambridge (1984).

Foucault, M. *The Order of Things*. In translation. Random House, New York (1970).

Frake, C. O. The ethnographic study of cognitive systems. In: Gladwin, T. and Sturdevant, W. C. *Anthropology and Human Behavior*. Anthropological Society of Washington, Washington, DC (1962).

Frake, C. O. Notes on queries in ethnography. In: Romney, A. K. and D'Andrade, R. G. eds. *Transcultural Studies in Cognition*. *American Anthropologist* 66(3):132–145, 1964. Part 2, special publication.

Frake, C. O. Plying frames can be dangerous: Some reflections on methodology in cognitive anthropology. *Quarterly Newsletter of the Institute for Comparative Human Development*, Rockefeller University 1(3):1–17, 1977.

Freeman, J. H. and Hannan, M. T. Growth and decline in organizations. *American Sociological Review* 40 (1975).

Freidson, E. Viewpoint: Sociology and medicine: A polemic. *Sociology of Health and Illness* 5(2):208–219, 1983.

Freud, S. The ego and the id. In: *The Standard Edition of the Complete Psychological Works of Sigmund Freud*. Vol. 19. (1974).

Freud, S. The question of lay analysis. *Standard Edition of the Complete Psychological Works of Sigmund Freud*. (1974).

Freud, S. *The Standard Edition of the Complete Psychological Works of Sigmund Freud*. Vol. 20. (1974).

Gadamer, H. G. The heritage of Hegel. *Reason in the Age of Science*. Lawrence, F. G. trans. MIT Press, Cambridge (1981).

Galilei, G., *Dialogues Concerning Two New Sciences*. Crew, H. and De Salvio, A. trans. Macmillan, New York (1914).

Garfinkel, H. *Studies in Ethnomethodology*. Prentice-Hall, Englewood Cliffs, NJ (1967).

Geertz, C. *The Interpretation of Cultures*. Basic Books, New York (1973).

Geertz, C. *Local Knowledge: Further Essays in Interpretive Anthropology*. Basic Books, New York (1983).

Geoghagen, W. H. *Decision-Making and Residence on Tagtabon Island*. Working paper No. 17, Language-Behavior Research Laboratory, University of California, Berkeley (1975).

Gerstein D. R. Introduction to: *Behavioral and Social Science: Fifty Years of Discovery*. National Academy Press, Washington (1986).

Ghiselli, E. E., Campbell, J. P. and Zedeck, S. *Measurement Theory for the Behavioral Sciences*. W. H. Freeman and Co., San Francisco (1981).

Gibbs, J. *Sociological Theory Construction*. Dryden, Hinsdale, IL. (1972).

Giddens, A. *The New Rules of Sociological Method*. Basic Books, New York (1976).

Gladwin, C. A model of the supply of smoked fish from Cape Coast to Kumasi. In: Plattner, S. ed. *Formal Methods in Economic Anthropology*, Special publication No. 4, American Anthropological Association, Washington, DC (1975).

Glaser, B. and Strauss, A. *The Discovery of Grounded Theory*. Aldine, Chicago (1967).

Glassner, B. and Loughlin, J. *Drugs in Adolescent Worlds: Burnouts to Straights*. Macmillan, London (1987).

Goodenough, W. Componential analysis and the study of meaning. *Language* **33**:195–216, 1956.

Goodenough, W. Cultural anthropology and linguistics. In: Hymes, D. ed. *Language in Culture and Society*. Harper and Row, New York (1964).

Goodenough, W. *Description and Comparison in Cultural Anthropology*. Aldine, Chicago (1970).

Goodenough, W. *Property, Kin, and Community on Truk,* Yale University publications in Anthropology No. 46. Yale University Press, New Haven (1951).

Goodman, N. *Fact, Fiction, and Forecast*. 3rd ed. Bobbs-Merrill, Indianapolis (1973).

Goodman, N. *Languages of Art*. Bobbs-Merrill, Indianapolis (1968).

Goodman, N. *Of Mind and Other Matters*. Harvard University Press, Cambridge (1984).

Goodman, N. On the statistical analysis of mobility tables. *American Journal of Sociology* **70** (1965).

Goodman, N. *The Structure of Appearance*. 3rd ed. Reidel, Dordrecht (1977).

Goodman, N. *Ways of Worldmaking*. Hackett Publishing Co., Indianapolis, (1978).

Goodman, N. The way the world is. In: *Problems and Projects*. Bobbs-Merrill, Indianapolis (1972).

Goodwin, P. Schopenhauer. *The Encyclopedia of Philosophy*. Vol. 8. Macmillan, New York (1967).

Gosnell, H. *Getting Out the Vote: An Experiment in the Stimulation of Voting*. University of Chicago Press, Chicago (1927).

Gosnell, H. *Machine Politics: Chicago Model.* University of Chicago Press, Chicago (1937).

Gould, L. C., Walker, A., Crane, L. and Lidz, C. W. *Connections: Notes from the Heroin World.* Yale University Press, New Haven, CT (1974).

Goulder, A. W. *The Two Marxisms.* Seabury Press, New York (1973).

Greenberg, J. H. An axiomatization of the phonological aspect of language. In: Gross, L. ed. *Symposium on Sociological Theory.* Harper and Row, New York (1959).

Gross, L. Theory construction in sociology: A methodological inquiry. In: Gross, L. ed. *Symposium on Sociological Theory.* Harper and Row, New York (1959).

Gubrium, J., Buckholdt, D. and Lynott, R. Considerations on a theory of descriptive activity. *Mid-American Review of Sociology* VII(1):17–35, 1982.

Gubrium, J. and Lynott, R. Family rhetoric as social order. *Journal of Family Issues* 6(1):129–52, 1985.

Gubrium, J. and Buckholdt, D. Fictive family: Everyday usage, analytic, and human service considerations. *American Anthropologist* 84(4):878–85, 1982.

Gumpertz, J. and Hymes, D. *Directions in Sociolinguistics: The Ethnography of Communication.* Holt, Rhinehart and Winston, New York (1972).

Guttenberg, M. and Secord, P. F. *Too Many Women? The Sex Ratio Question.* Sage, Beverly Hills (1983).

Gutting, G. ed. *Paradigms and Revolutions.* University of Notre Dame Press, Notre Dame, (1980).

Habermas, J. *Knowledge and Human Interests.* Beacon Press, Boston (1972).

Habermas, J. *Knowledge and Human Interests.* Shapiro, J. J. trans. Beacon Press, Boston (1971).

Habermas, J. What is universal pragmatics? *Communication and Evolution of Society.* McCarthy, T. trans. Beacon Press, Boston (1979).

Hacking, I. *Why Does Language Matter to Philosophy?* Cambridge University Press, Cambridge (1975).

Hagerstrand, T. The Monte-Carlo simulation of diffusion. In: Garrison, W. L. and Marble, D. F. eds. *Quantitative Geography.* Northwestern University Press, Evanston (1960).

Halbwachs, M. *The Causes of Suicide.* Routledge and Kegan Paul, London (1930, 1978).

Halbwachs, M. *The Collective Memory.* Harper and Row, New York (1950, 1980).

Hamblin, R. L. *et al. A Mathematical Theory of Social Change.* Wiley, New York (1973).

Hamilton, J. D. and Hamilton, L. C. Models of social contagion. *Journal of Mathematical Society* 8 (1981).

Hamlyn, D. W. *Schopenhauer.* Routledge and Kegan Paul, London (1980).

Hammersley, M. From ethnography to theory: A programme and paradigm in the sociology of education. *Sociology* 19(2):244–59, 1985.

Hammersley, M. and Atkinson, P. *Ethnography: Principles in Practice.* Tavistock, London (1983).

Harré, R. *Principles of Scientific Thinking.* University of Chicago Press, Chicago (1970).

Harré, R. *Varieties of Realism.* Basil Blackwell, New York and Oxford (1986).

Harré, R. and Madden, E. *Casual Powers: A Theory of Natural Necessity.* Rowan and Littlefield, Totowa, NJ (1975).

Harris, M. *Cultural Materialism: The Struggle for a Science of Culture.* Random House, New York (1979).

Harris, M. History and significance of the emic-etic distinction. In: Siegel *et al.* eds. *Annual Review of Anthropology* 5:329–350. Annual Reviews, Inc, Palo Alto (1976).

Harris, M. *The Rise of Anthropological Theory.* Crowell, New York (1968).

Hart, H. L. A. *The Concept of Law.* Oxford University Press, Oxford (1961).

Heath, C. Participation in the medical consultation: The coordination of verbal and non-verbal behavior between doctor and patient. *Sociology of Health and Illness* 6(3):311–38, 1984.

Hegel, G. W. F. *Hegel's Logic, Being Part I of the Encyclopedia of the Philosophical Sciences (1830) translated by William Wallace.* Clarendon Press, Oxford (1975).

Hempel, C. G. *Aspects of Scientific Explanation.* Free Press, New York (1965).

Hempel, C. G. Logical positivism and the social sciences. In: Achinstein, P. and Barker, S. F. eds. *The Legacy of Logical Positivism.* Johns Hopkins University Press, Baltimore (1969).

Hempel, C. G. *Philosophy of Natural Science.* Prentice-Hall, Englewood Cliffs (1966).

Hirsch, J. Behavior-genetic analysis and its biosocial consequences. In: Dworkin, G. ed. *The IQ Controversy.* Pantheon, New York (1976).

Hockney, D. The bifurcation of scientific theories and indeterminacy of translation. *Philosophy of Science* **XLII**, 1975.

Hollis, M. The epistemological unity of mankind. In: Brown, S. C. ed. *Philosophical Disputes in the Social Sciences.* Harvester, Brighton (1979).

Hollis, M. The limits of irrationality. In: Wilson, B. ed. *Rationality.* Blackwell, Oxford (1970).

Hollis, M. Reason and ritual. In: Wilson, B. ed. *Rationality.* Blackwell, Oxford (1970).

Hollis, M. and Lukes, S. eds. *Rationality and Relativism.* MIT Press, Cambridge (1982).

Horowitz, I. L. Disenthralling sociology. *Society* 24:48–55, 1987.

Huberman, A. and Miles, M. Drawing valid meaning from qualitative data: Some techniques of data reduction and display. *Quality and Quantity* **17**:281–339, 1983.

Husserl, E. *The Crisis of European Sciences and Transcendental Phenomenological Philosophy.* Carr, D. trans. Northwestern University Press, Evanston (1970).

Hymes, D. H. Linguistic method in ethnography: Its development in the United States. In: Garvin, P. L. ed. *Method and Theory in Linguistics.* Mouton, The Hague (1970).

Janik, A. and Toulmin, S. *Wittgenstein's Vienna.* Simon and Schuster, New York (1973).

Johnson, J. *Doing Field Research.* Free Press, New York (1976).

Johnston, D. F. ed. *Measurement of Subjective Phenomena.* U.S. Department of Commerce (Special Demographic Analyses), Washington (1981).

Jones, E. *The Life and Work of Sigmund Freud.* Vol. 1–3. Basic Books, New York (1981).

Jorion, P. Emic and etic: Two anthropological ways of spilling ink. *Cambridge Anthropology* **8**:41–68, 1983.

Kant, I. *Critique of Pure Reason*. Smith, N. K. trans. Macmillan, New York (1958).

Kaplan, A. Measurement in behavioral sciences. In: Brodbeck, M. ed. *Readings in the Philosophy of the Social Sciences*. Macmillan, New York, (1968).

Karl, B. *Charles E. Merriam and the Study of Politics*. University of Chicago Press, Chicago (1974).

Kay, P. Some theoretical implications of ethnographic semantics. *Current Directions in Anthropology*, American Anthropological Association Bulletin **3**(2):Part 2, 1970.

Kerlinger, F. N. *Behavioral Research: A Conceptual Approach*. Holt, Rinehart and Winston, New York (1979).

Kershaw, D. and Fair, J. *The New Jersey Income-Maintenance Experiment*. Three vols. Academic Press, New York (1976–77).

Kiefer, C. Psychological anthropology. In: Siegel *et al.* eds. *Annual Review of Anthropology* **6**:103–119. Annual Reviews, Inc., Palo Alto (1977).

Kiernan, C. *The Enlightenment and Science in Eighteenth Century France*. Institut et Musée Voltaire, Geneva, (1968). 2nd ed. Cheney and Sons, Banbury, Oxfordshire, (1973).

Klerman, G. ed. *Suicide and Depression among Adolescents and Young Adults*. American Psychiatric Association, Washington (1986).

Kripke, S. *Wittgenstein on Rules and Private Language*. Harvard University Press, Cambridge (1982).

Kuhn, T. S. Postscript 1969. *The Structure of Scientific Revolutions*. 2nd ed. enl. University of Chicago Press, Chicago, (1962, 1970).

Kuhn, T. S. Reflections on my critics. In: Lakatos, I. and Musgrave, A. eds. *Criticism and the Growth of Knowledge*. Cambridge University Press, Cambridge, (1970).

Kuhn, T. S. Second thoughts on paradigms. In: Suppes, F. ed. *The Structure of Scientific Theories*. University of Illinois Press, Urbana, (1970).

Kuhn, T. S. *Structure of Scientific Revolutions*. 2nd ed. enl. University of Chicago Press, Chicago, (1962, 1970).

Kweit, M. and Kweit, R. *Concepts and Methods for Political Analysis*. Prentice-Hall, Englewood Cliffs (1981).

Lacan, J. *Speech and Language in Psychoanalysis*. Wilden, A. trans. Johns Hopkins University Press, Baltimore (1968). Orig. pub. as *Language of the Self*.

Lakatos, I. Falsification and the methodology of scientific research programmes. Worrall, J. and Currie, G. eds. *Philosophical Papers*. Vol. 1. Cambridge University Press, Cambridge, (1978). Orig. publ. in: Lakatos, I. and Musgrave, A. eds. *Criticism and the Growth of Knowledge*. Cambridge University Press, Cambridge, (1970).

Lalande, A. Allocution pour le centenaire de la naissance d'Emile Durkheim. *Annales de l'Université de Paris*. (1960).

Lalande, A. *Vocabulaire technique et critique de la philosophie*. Presses Universitaires de France, Paris (1926, 1980).

Larson, D. Correspondence and the third dogma. *Dialectica* **41**:231–237, 1987.

Laski, H. *On the Study of Politics: An Inaugural Lecture Delivered at the London School of Economics and Political Science on 22 October 1926*. Humphrey

Milford/Oxford University Press, Oxford (1926).

Lasswell, H. The cross-disciplinary manifold: The Chicago prototype. In: Lepawsky, A., Buehrig, E., and Lasswell, H. eds. *The Search for World Order: Studies by Students and Colleagues of Quincy Wright.* Appleton-Century-Crofts, New York (1971).

Lasswell, H. General framework: Person, personality, group, culture. In: *The Analysis of Political Behavior: An Empirical Approach.* Oxford University Press, New York (1948). Orig. publ. *Psychiatry* 2:375–390, 1939.

Lasswell, H. *The Policy Sciences.* Stanford University Press, Stanford (1959).

Lasswell, H. *Politics: Who Gets What, When, How.* McGraw-Hill, New York (1936).

Lasswell, H. *Propaganda Techniques in the World War.* Alfred A. Knopf, New York (1927).

Lasswell, H. *Psychopathology and Politics.* In: *The Political Writings of Harold D. Lasswell.* Free Press, Glencoe (1951). Orig. publ. University of Chicago Press, Chicago (1930).

Lasswell, H. The qualitative and quantitative in political and legal analysis. In: Lerner, D. ed. *Quantity and Quality.* Free Press, Glencoe (1961).

Lasswell, H. Why be quantitative? In: Lasswell, H., Leites, N. *et al. Language of Politics: Studies in Quantitative Semantics.* George W. Stewart, New York (1949).

Lasswell, H. *World Politics and Personal Insecurity.* Whittlesey House, New York (1935).

Lasswell, H. and Kaplan, A. *Power and Society: A Framework for Political Inquiry.* Yale University Press, New Haven (1950).

Lazarsfeld, P. Interpretation of statistical relations as research operation. In: Lazarsfeld, P. F. and Ginsber, M. eds. *The Language of Social Research.* Free Press, New York (1955).

Lazarsfeld, P. Preface to: Oberschall, A. *Empirical Research in Germany, 1848–1914.* Basic Books, New York (1965).

Lear, J. Leaving the world alone. *The Journal of Philosophy* 79:382–403, 1982.

Lerner, D. ed. *Quantity and Quality.* Free Press, New York (1961).

Lévi-Strauss, C. *The Raw and Cooked: Introduction to a Science of Mythology.* Weightman, J. and D. trans. Harper and Row, New York (1969).

Lévi-Strauss, C. *The Savage Mind.* in trans. University of Chicago Press, Chicago (1966).

Lévi-Strauss, C. The structural study of myth. *Journal of American Folklore* 78(270):428–44, 1955.

Levy-Bruhl, L. *The History of Philosophy in France.* Open Court, Chicago (1899).

Lewontin, R. C. The analysis of variance and the analysis of causes. In: Dworkin, G. ed. *The IQ Controversy.* Pantheon, New York (1976).

Lidz, C. W. The cop-addict game: A model of police suspect-interaction. *Journal of Political Science Administration* 2(1), 1974.

Lukes, S. *Emile Durkheim: His Life and Work.* Harper and Row, New York (1972).

Mach, E. *Popular Scientific Lectures.* 5th ed. Open Court, Lasalle (1945).

MacIntyre, A. Is a science of comparative politics possible? In: *Against the Self-Images of the Age.* University of Notre Dame Press, Notre Dame (1971).

MacKenzie, D. A. *Statistics in Britain 1865–1930.* Edinburgh University Press, Edinburgh (1981).

Magree, B. *The Philosophy of Schopenhauer.* Oxford University Press, New York (1983).

Makkreel, R. A. *Dilthey: Philosopher of the Human Sciences.* Princeton University Press, Princeton (1975).

Manheim, J. and Rich, R. *Empirical Political Analysis: Research Methods in Political Science.* Prentice-Hall, Englewood Cliffs (1981).

Manicas, P. T. *History and Philosophy of the Sciences.* Basil Blackwell, New York and Oxford (1987).

Mann, T. Introduction. In: Durant, W. ed. *The Works of Schopenhauer.* Federick Unger Publishers, New York (1939, 1955).

Marano, L. Windigo psychosis: The anatomy of an emic-etic confusion. *Current Anthropology* 23:385–412, 1982.

Margolis, J. *Culture and Cultural Entities.* D. Reidel, Dordrecht (forthcoming).

Margolis, J. Goethe and psychoanalysis. In: Zucker, F. J., Amrine, F. and Wheeler, H. eds. *Goethe and the Sciences: A Reappraisal.* D. Reidel, Dordrecht (1986).

Margolis, J. Historicism, universalism and the threat of relativism. Unpublished.

Margolis, J. The nature of strategies of relativism. *Mind* XCII, 1983.

Margolis, J. *Persons and Minds* D. Reidel, Dordrecht (1978).

Margolis, J. *Philosophy of Psychology.* Prentice-Hall, Englewood Cliffs (1984).

Margolis, J. Pragmatism without foundations. *American Philosophical Quarterly* XXI, 1984.

Margolis, J. Reconciling Freud's *Scientific Project* and psychoanalysis. In: Callahan, D. and Engelhardt, H. T., Jr. eds. *Morals, Science and Sociality* (Vol. III) *The Foundations of Ethics and Its Relationship to Science.* The Hastings Center, Hastings-on-Hudson (1978).

Margolis, J. Relativism, history, and objectivity in the human studies. *Journal for the Theory of Social Behavior* XIV (1984).

Margolis, J. The savage mind totalizes. *Man and World* XVII, 1984.

Margolis, J. Schlick and Carnap on the problem of psychology. In: Gadol, E. T. ed. *Rationality and Science.* Springer-Verlag, Vienna (1982).

Margolis, J. Scientific realism as a transcendental issue. *Manuscito* VII (1984).

Margolis, J. Skepticism, foundationalism and pragmatism. *American Philosophical Quarterly* XIV, 1977.

Marjoribanks, K. Ethnic and environmental influences on mental abilities. *American Journal of Sociology* 78 (1972).

Marsh, C. Problems with surveys: Method and epistemology. *Sociology* 13 (2):293–305, 1979.

Marshall, C. Appropriate criteria of the trustworthiness and goodness for qualitative research on educational organizations. *Quality and Quantity* 19:353–73, 1985.

Martindale, D. *The Nature and Types of Sociological Theory.* 2nd ed. Houghton Mifflin, Boston (1981).

Marvick, D. Introduction: Context, problems, and methods. In: Marvick, D. ed. *Harold D. Lasswell on Political Sociology.* University of Chicago Press, Chicago (1977).

Masterman, M. The nature of a paradigm. In: Lakatos, I. and Musgrave, A. eds. *Criticism and the Growth of Knowledge.* Cambridge University Press, Cambridge, (1970).

Mauss, M. *The Gift*. Free Press, Glencoe, IL (1954).

Mauss, M. *Sociology and Psychology*. Routledge and Kegan Paul, London (1950, 1979).

Mayberry-Lewis, D. *The Savage and the Innocent*. Beacon Press, Boston (1965).

Mayhew, D. *Congress: The Electoral Connection*. Yale University Press, New Haven (1974).

Mayr, E. *The Growth of Biological Thought*. Harvard University Press, Cambridge (1982).

McCarthy, T. Rationality and relativism: Habermas's 'Overcoming' of hermeneutics. In: Thompson, J. B. and Held, D. eds. *Habermas: Critical Debates*. MIT Press, Cambridge (1982).

McGinnis, R. A stochastic model of social mobility. *American Sociological Review* 33 (1968).

Mead, G. H. *Mind, Self and Society*. University of Chicago Press, Chicago (1934).

Mead, G. H. *Mind, Self and Society from the Standpoint of a Social Behaviorist*. Morris, C. ed. University of Chicago Press, Chicago (1934).

Merriam, C. *American Political Ideas: Studies in the Development of American Political Thought, 1865–1917*. Macmillan, New York (1920).

Merriam, C. *The American Party System: An Introduction to the Study of Political Parties in the United States*. Macmillan, New York (1922).

Merriam, C. The education of Charles E. Merriam. In: White, L. ed. *The Future of Government in the United States: Essays in Honor of Charles E. Merriam*. University of Chicago Press, Chicago (1942).

Merriam, C. *New Aspects of Politics*. University of Chicago Press, Chicago (1925).

Merriam, C. *Political Power: Its Composition and Incidence*. Whittlesey House, New York (1934).

Merriam, C. and Gosnell, H. *The American Party System: An Introduction to the Study of Political Parties in the United States*. rev. ed. Macmillan, New York (1929).

Merriam, C. and Gosnell, H. *The American Party System: An Introduction to the Study of Political Parties in the United States*. 3rd ed. Macmillan, New York (1940).

Merriam, C. and Gosnell, H. *The American Party System: An Introduction to the Study of Political Parties in the United States*. 4th ed. Macmillan, New York (1949).

Merriam, C. and Gosnell, H. *Non-voting: Causes and Methods of Control*. University of Chicago Press, Chicago (1924).

Merton, R. K. The bearing of empirical research on sociological theory. In: *Social Theory and Social Structure*. rev. Free Press, New York (1957).

Merton, R. K. *Social Theory and Social Structure*. Free Press, New York (1957).

Meštrović, S. G. Durkheim, Schopenhauer and the relationship between goals and means: Reversing the assumptions in the parsonian theory of rational action. *Sociological Inquiry*. Forthcoming (1988).

Meštrović, S. G. Durkheim's concept of anomie considered as a 'total' social fact. *British Journal of Sociology* 38(4):567–583, 1987.

Meštrović, S. G. Durkheim's concept of the unconscious. *Current Perspectives in Social Theory* 5:267–88, 1984.

Meštrović, S. G. Durkheim's conceptualization of political anomie. *Research in Political Sociology*. Forthcoming (1988).

Meštrović, S. G. Durkheim's renovated rationalism and the idea that 'collective life is only made of representations'. *Current Perspectives in Social Theory* **6**:199–218, 1985.

Meštrović, S. G. *In the Shadow of Plato: Durkheim and Freud on Suicide and Society*. Doctoral dissertation, Syracuse University (1982).

Meštrović, S. G. Simmel's concept of the unconscious. Paper presented to the Western Social Science Association, San Diego (1984).

Meštrović, S. G. A sociological conceptualization of trauma. *Social Science and Medicine* **21**: 835–48, 1985.

Meštrović, S. G. and Brown, H. M. Durkheim's concept of anomie as dereglement. *Social Problems* **33**: 81–99, 19.

Meštrović, S. G. and Glassner, B. A Durkheimian hypothesis on stress. *Social Science and Medicine* **17**: 1315–27, 1983.

Miller, S. M. The Participant observer and 'overrapport.' *American Sociology Review* **18**: 97–99, 1953.

Mills, C. W. Situated actions and vocabularies of motive. *American Sociological Review* **5**: 904–13, 1940.

Mitchell, J. C. Case and situational analysis. *Sociological Review* **31**(2):187–211, 1983.

Mizrahi, T. Getting rid of patients. *Sociology of Health and Illness* **7**(2):214–235, 1985.

Nagel, E. Measurement. In: Danto, A. and Morgenbesser, S. *Philosophy of Science*. Meridian, Cleveland, (1967).

Neurath, O. Protocol sentences. Schick, G. trans. In: Ayer, A. J. ed. *Logical Positivism*. Free Press, Glencoe (1959).

Oberschall, A. ed. *The Establishment of Empirical Sociology*. Harper and Row, New York (1972).

Oppenheim, P. and Putnam, H. Unity of science as a working hypothesis. In: Feigl, H. *et al.* eds. *Minnesota Studies in the Philosophy of Science*. Vol I. University of Minnesota Press, Minneapolis (1958).

Pareto, W. *Manual of Political Economy*. Kelley, New York (1971).

Pareto, W. *Mind and Society*. Harcourt, New York (1935).

Park, R. The city: Suggestions for the investigation of human behavior in the urban environment. In: Park, R., Burgess, E., and MacKenzie, R. *The City*. University of Chicago Press, Chicago (1925).

Parsons, T. Evaluation and objectivity in the social sciences: An interpretation of Max Weber's contributions. *Deutsche Gesellschaft für Soziologie and the UNESCO Journal of the Social Sciences* **17**(1), 1975.

Petrie, A. *Individuality in Pain and Suffering*. University of Chicago Press, Chicago (1967).

Piaget, J. *Biology and Knowledge*. University of Chicago Press, Chicago (1971).

Piaget, J. *The Child's Conception of the World*. Littlefield, Adam, Paterson, NJ (1960).

Piaget, J. Piaget's theory. In: Inhelder, B. *et al.* eds. *Piaget and His School*. Springer-Verlag, New York (1976). Orig. in: Mussen, P. H. ed. *Carmichael's Manual of*

Child Psychology. 3rd ed. Vol 1. John Wiley and Sons, New York (1970).

Piaget, J. *Structuralism*. Maschler, C. trans. Basic Books, New York (1970).

Piattelli-Palmarini, M. ed. *Language and Learning: The Debate Between Jean Piaget and Noam Chomsky*. Harvard University Press, Cambridge (1980).

Pike, N. *Language in Relation to United Theory of the Structure of Human Behavior: Part I*. Summer Institute of Linguistics, Glendale, CA (1954).

Pike, N. Towards a theory of the structure of human behavior. In: *Estudios Antropologicos Publicados en Homenaje al Doctor Manuel Gamio*. Sociedad Mexicana de Antropologia, Mexica, DF (1956).

Pitkin, H. *Wittgenstein and Justice*. University of California Press, Berkeley (1972).

Plato. *Republic*. Cornford, F. M. trans. Oxford University Press, Oxford, (1972).

Platt, J. The social construction of 'positivism' and its significance in British sociology: 1950–80. In: Abrams, P. *et al. Practice and Progress: British Sociology 1950–80*. Allen and Unwin, London (1981).

Polanyi, L. *Telling the American Story: A Structural and Cultural Analysis of Conversational Storytelling*. Ablex, Norwood (1985).

Popper, K. R. The aim of science. *Objective Knowledge*. Clarendon Press, Oxford (1972).

Popper, K. R. *Conjectures and Refutations*. 2nd ed. Routledge and Kegan Paul, London (1965).

Popper, K. R. Normal science and its dangers. In: Lakatos, I. and Musgrave, A. eds. *Criticism and the Growth of Knowledge*. Cambridge University Press, Cambridge, (1970).

Popper, K. R. *Realism and the Aim of Science (Postscript to the Logic of Scientific Discovery*, Vol I.) Bartley, WW, III. ed. Rowman and Littlefield, Totowa, NJ (1983).

Porpora, D. A. *The Concept of Social Structure*. Greenwood Press, Denver (1987).

Post, E. *Etiquette*. Funk and Wagnells, New York (1945).

Prior, L. Making sense of mortality. *Sociology of Health and Illness* 7(2):167–190, 1985.

Putnam, H. *Reason, Truth and History*. Cambridge University Press, Cambridge (1981).

Putnam, H. *Meaning and the Moral Sciences*. Routledge and Kegan Paul, London (1978).

Quine, W. V. Epistemology naturalized. In: *Ontological Relativity and Other Essays*. Columbia, New York (1960).

Quine, W. V. Facts of the matter. In: Shahan, R. W. and Swoyer, C. *Essays on the Philosophy of W. V. Quine*. University of Oklahoma, Norman (1979).

Quine, W. V. Goodman's *Ways of Worldmaking*. In: *Theories and Things*. Harvard, Cambridge (1981).

Quine, W. V. *Meaning and Object*. MIT Press, Cambridge (1960).

Quine, W. V. On the reasons for indeterminacy of translation. *The Journal of Philosophy* **67**:179:183, 1970.

Quine, W. V. The problem of meaning in linguistics. In: *From a Logical Point of View*. 2nd ed. Harper and Row, New York (1951, 1961).

Quine, W. V. Reply to Paul A. Roth. In: Hahn, L. and Schilpp, P. eds. *The Philosophy of W. V. Quine*. Open Court Press, (1987).

Quine, W. V. Two dogmas of empiricism. In: *From a Logical Point of View*. 2nd ed. Harper and Row, New York (1951, 1961).

Quine, W. V. *Word and Object*. MIT Press, Cambridge (1960).

Quinn, N. Do Mfantse fish sellers estimate probabilities in their heads? *American Ethnologist* 5:206–226, 1978.

Rawls, J. *A Theory of Justice*. Harvard, Cambridge (1971).

Reason, D. Mathematical models and understanding social life. Unpublished paper. University of Kent at Canterbury, nd.

Redfield, R. *Tepoztlan, A Mexican Village: A Study of Folk Life*. University of Chicago Press, Chicago (1930).

Reichenbach, H. *Laws, Modalities and Counterfactuals*. University of California Press, Berkeley (1976).

Reid, T. *The Works*. Maclachlan, Steward and Company, Edinburgh, (1849).

Rice, S. *Quantitative Methods in Politics*. Alfred A. Knopf, New York (1928).

Ricoeur, P. *Hermeneutics and the Human Sciences*. Thompson, J. B. ed. and trans. Cambridge University Press, Cambridge (1981).

Roadburg, A. Breaking relationships with research subjects: Some problems and suggestions. In: Shaffir, R., Stebbins, R. and Turowetz, A. *Field Work Experience: Approaches to Social Research*. St. Martin's Press, New York (1980).

Rorty, R. *Consequences of Pragmatism*. University of Minnesota Press, Minneapolis (1982).

Rorty, R. Method, social science, and social hope. In: *Consequences of Pragmatism*. University of Minnesota Press, Minneapolis (1982).

Rorty, R. *Philosophy and the Mirror of Nature*. Princeton University Press, Princeton (1979).

Rossi, P. H. and Lyall, K. C. *Reforming Public Welfare: A Critique of the Negative Income Tax Experiment*. Russell Sage Foundation, New York (1976).

Rosten, L. Harold Lasswell: A memoir. In: Rogow, A. *Politics, Personality, and Social Science in the Twentieth Century: Essays in Honor of Harold D. Lasswell*. University of Chicago Press, Chicago (1969).

Roszak, T. *The Making of a Counter Culture*. Doubleday, Garden City, NY, (1969).

Rudner, R. *Philosophy of the Social Sciences*. Prentice-Hall, Englewood Cliffs (1966).

Rudner, R. Some essays at objectivity. *Philosophical Exchange* 1:115–135, 1973.

Russell, B. On the notion of cause, with application to the free-will problem. In: Feigl, H. and Brodbeck, M. eds. *Readings in the Philosophy of Science*. University of California Press, Berkeley (1951).

Sacks, O. *The Man Who Mistook His Wife for a Hat, and Other Clinical Tales*. Summit Books, New York (1985).

Salmon, W. C. Statistical explanation. In: Salmon, W. C. *et al. Statistical Explanation and Statistical Relevance*. University of Pittsburgh Press, Pittsburgh (1971).

Salmon, W. C. Why ask 'why'? Presidential address, *American Philosophical Association* 51 (1978).

Sampson, R. V. *Progress in the Age of Reason*. Harvard University Press, Cambridge, MA, (1956).

Sartori, G., Riggs, F., and Teune, H. Tower of Babel: On the definition and analysis of concepts in social sciences. *Occasional Paper* 6. International Studies Association, Pittsburgh (1975).

Schopenhauer, A. *The World as Will and Idea*. AMS Press, New York (1818, 1977).

Schuman, F. L. *International Politics: An Introduction to the Western State System*. McGraw-Hill, New York (1933).

Schumpeter, J. In: Schumpeter, E. ed. *History of Economic Analysis*. Oxford University Press, New York (1954).

Schutz, N. On the anatomy and comparability of linguistics and ethnographic description: Toward a generative theory of ethnography. In: Kinkade, M. D. *et al.* eds. *Linguistics and Anthropology: In Honor of C. F. Voegelin*. De Ridder, Lisse (1975).

Schwartz, H. and Jacobs, J. *Qualitative Sociology: A Method to the Madness*. Free Press, New York (1979).

Schwartz, M. and Schwartz, C. G. Problems in participant observation. *American Journal of Sociology* 60(4):343–353, 1955.

Scott, J. *The Moral Economy of the Peasant: Rebellion and Subsistence in Southeast Asia*. Yale University Press, New Haven (1976).

Scott, J. *Weapons of the Weak: Everyday Forms of Peasant Resistance*. Yale University Press, New Haven (1985).

Seidel, J. and Clark, J. The ethnograph. *Qualitative Sociology* 7(1&2):110–125, 1984.

Selvin, H. On following in someone's footsteps': Two examples of Lazarsfeldian methodology. In: Merton, R. K., Coleman, J. S., and Rossi, P. S. eds. *Qualitative and Quantitative Social Research: Papers in Honor of Paul S. Lazarsfeld*. The Free Press, New York (1979).

Shapere, D. Meaning and scientific change. In: Colodny, R. G. ed. *Mind and Cosmos: Essays in Contemporary Science and Philosophy*. University of Pittsburgh Press, Pittsburgh (1966).

Shapere, D. The structure of scientific revolutions. *Philosophical Review* LXXIII, 1964.

Silverman, D. The clinical subject: Adolescents in a cleft-palate clinic. *Sociology of Health and Illness* 5(3):253–74, 1983.

Silverman, D. *Communication and Medical Practice: Social Relations in the Clinic*. Sage, London and Beverly Hills (1987).

Silverman, D. Going private: Ceremonial forms in a private oncology clinic. *Sociology of Health and Illness* 18(4):191–202, 1984.

Silverman, D. *Qualitative Methodology and Sociology*. Gower, Aldershot (1985).

Silverman, D. Six rules of qualitative research: A post-romantic argument. *Symbolic Interaction* (forthcoming).

Silverman, D. and Torode, B. *The Material Word: Some Theories of Language and Its Limits*. Routledge, London (1980).

Simmel, G. *George Simmel on Individuality and Its Social Forms*. University of Chicago Press, Chicago (1971).

Simmel, G. *Schopenhauer and Nietzsche*. University of Massachusetts Press, Amherst, (1907, 1986).

Simmel, G. *The Sociology of Georg Simmel*. Wolff, K. H. ed. Free Press, Glencoe, IL (1950).

Simmel, G. *The Sociology of Georg Simmel*. Wolff, K. H. ed. and trans. Free Press, New York (1964).

Smelser, N. J. The Ogburn vision fifty years later. In: *Behavioral and Social Science, Fifty Years of Discovery*. National Academy Press, Washington (1986).

Smith, B. The mystifying intellectual history of Harold D. Lasswell. In: Rogow, A. *Politics, Personality, and Social Science in the Twentieth Century: Essays in Honor of Harold D. Lasswell*. University of Chicago Press, Chicago (1969).

Smith, C. W. On the sociology of the mind. In: Secord, P. *Explaining Human Behavior*. Sage Publication, Beverly Hills (1982).

Smith, H. W. *Strategies of Social Research: The Methodological Imagination*. Prentice-Hall, London (1975).

Smith, R. B. *An Introduction to Social Research*. Ballinger, Cambridge, MA (1983).

Somit, A. and Tanenhaus, J. *The Development of American Political Science: From Burgess to Behavioralism*. Allyn and Bacon, Boston (1967).

Spradley, J. *The Ethnographic Interview*. Holt, Rinehart and Winston, New York (1979).

Strawson, P. *Individuals*. Methuen, London (1959).

Strong, P. *The Ceremonial Order of the Clinic*. Routledge and Kegan Paul, London (1979).

Strong, P. Seminar paper, Department of Sociology, Goldsmiths' College (1985).

Stroud, W. L. Biographical explanation is low-powered science. *American Psychologist* 39 (1984).

Sudnow, D. *Passing On: The Social Organization of Dying*. Prentice-Hall, Englewood Cliffs, NJ (1967).

Sylvan, D. and Glassner, B. *A Rationalist Methodology for the Social Sciences*. Basil Blackwell, London, New York and Oxford (1985).

Taylor, C. Interpretation and the sciences of man. *Review of Metaphysics* 25:3–51, 1971.

Tönnies, F. *Community and Society*. Harper and Row, New York (1887, 1963).

Toulmin, S. Does the distinction between normal and revolutionary science hold water? In: Lakatos, I. and Musgrave, A. eds. *Criticism and the Growth of Knowledge*. Cambridge University Press, Cambridge, (1970).

Triandis, H. C. Approaches toward minimizing translation. In: Brislin, R. W. ed. *Translation: Application and Research*. Wiley, New York (1976).

Trigg, R. *Understanding Social Science*. Basil Blackwell, New York (1985).

Vygotsky, L. S. *Thought and Language*. Hanfmann, E. and Vakar, G. trans. MIT Press, Cambridge (1962).

Watson, L. C. 'Etic' and 'emic' perspectives on Guajiro urbanization. *Urban Life* 9:441–468, 1981.

Wax, R. Field methods techniques: Reciprocity as a field technique. *Human Organization* 11(3):34–37, 1952.

Weber, M. *On the Methodology of the Social Sciences*. Free Press, Glencoe, IL (1949).

Weber, M. *The Protestant Ethic and the Spirit of Capitalism*. Charles Scribner's Sons, New York (1904, 1958).

Werner, O. Ethnoscience 1972. In: Siegel *et al*. eds. *Annual Reviews in Anthropology* 1:271–308. Annual Reviews, Inc., Palo Alto (1972).

Wesley, C. S. *Statistical Explanation and Statistical Relevance*. University of Pittsburgh Press, Pittsburgh (1971).

White, L. *The Prestige Value of Public Employment in Chicago.* University of Chicago Press, Chicago (1929).

Williams, G. The genesis of chronic Illness: Narrative reconstruction. *Sociology of Health Illness* 6(2):175–200, 1984.

Winch, P. *The Idea of a Social Science and Its Relation to Philosophy.* Routledge and Kegan Paul, London (1958).

Wirth, L. *The Ghetto.* University of Chicago Press, Chicago (1928).

Wirth, L. Preface to: Wirth, L. and Shils, E. trans. *Ideology and Utopia.* Harcourt, Brace, New York (1936).

Wirth, L. ed. *Eleven Twenty-Six: A Decade of Social Science Research.* University of Chicago Press, Chicago (1940).

Wright, Q. *A Study of War.* 2nd. edn. University of Chicago Press, Chicago (1965). Orig. publ. (1942).

Wrong, D. The over-socialized conception of man in modern sociology. *American Sociological Review* 26:183–93, 1961.

Wundt, W. *Ethics.* Macmillan, New York (1886, 1902).

Zerubavel, E. *Patterns of Time.* University of Chicago Press, Chicago (1979).

LIST OF CONTRIBUTORS

PETER CAWS is University Professor of Philosophy, George Washington University, Washington, DC.

ROBERT FELEPPA is Associate Professor of Philosophy, Wichita State University, Wichita, Kansas.

CHARLES W. LIDZ is Professor of Psychiatry and Sociology, University of Pittsburgh, Pittsburgh, Pennsylvania.

PETER T. MANICAS is Professor of Philosophy, Queens College of the City University of New York, New York.

JOSEPH MARGOLIS is Professor of Philosophy, Temple University, Philadelphia, Pennsylvania.

STJEPAN G. MEŠTROVIĆ is Associate Professor of Sociology, Lander College, Greenwood, South Carolina.

DAVID SILVERMAN is Reader in Sociology, Goldsmiths' College, University of London, England.

CHARLES W. SMITH is Professor and Chair of Sociology, Queens College of the City University of New York, New York.

DAVID J. SYLVAN is Associate Professor of Political Science, University of Minnesota, Minneapolis, Minnesota.

BARRY GLASSNER is Professor and Head of Sociology, University of Connecticut, Storrs, Connecticut.

JONATHAN D. MORENO is Director of the Division of Humanities in Medicine and Professor, State University of New York Health Science Center at Brooklyn, New York.

INDEX OF NAMES

BOSTON STUDIES IN THE PHILOSOPHY OF SCIENCE

Editors:

ROBERT S. COHEN and MARX W. WARTOFSKY
(Boston University)

20. Kenneth F. Schaffner and Robert S. Cohen (eds.), *Proceedings of the 1972 Biennial Meeting, Philosophy of Science Association.* 1974
21. R. S. Cohen and J. J. Stachel (eds.), *Selected Papers of Léon Rosenfeld.* 1978.
22. Milic Čapek (ed.), *The Concepts of Space and Time. Their Structure and Their Development.* 1976.
23. Marjorie Grene, *The Understanding of Nature, Essays in the Philosophy of Biology.* 1974.
24. Don Ihde, *Technics and Praxis. A Philosophy of Technology.* 1978.
25. Jaakko Hintikka and Unto Remes, *The Method of Analysis. Its Geometrical Origin and Its General Significance.* 1974.
26. John Emery Murdoch and Edith Dudley Sylla, *The Cultural Context of Medieval Learning.* 1975.
27. Marjorie Grene and Everett Mendelsohn (eds.), *Topics in the Philosophy of Biology.* 1976.
28. Joseph Agassi, *Science in Flux.* 1975.
29. Jerzy J. Wiatr (ed.), *Polish Essays in the Methodology of the Social Sciences.* 1979.
30. Peter Janich, *Protophysics of Time.* 1985.
31. Robert S. Cohen and Marx W. Wartofsky (eds.), *Language, Logic and Method.* 1983.
32. R. S. Cohen, C. A. Hooker, A. C. Michalos, and J. W. van Evra (eds.), *PSA 1974: Proceedings of the 1974 Biennial Meeting of the Philosophy of Science Association.* 1976.
33. Gerald Holton and William Blanpied (eds.), *Science and Its Public: The Changing Relationship.* 1976.
34. Mirko D. Grmek (ed.), *On Scientific Discovery.* 1980.
35. Stefan Amsterdamski, *Between Experience and Metaphysics. Philosophical Problems of the Evolution of Science.* 1975.
36. Mihailo Marković and Gajo Petrović (eds.), *Praxis, Yugoslav Essays in the Philosophy and Methodology of the Social Sciences.* 1979.
37. Hermann von Helmholtz, *Epistemological Writings. The Paul Hertz/Moritz Schlick Centenary Edition of 1921 with Notes and Commentary by the Editors.* (Newly translated by Malcolm F. Lowe. Edited, with an Introduction and Bibliography, by Robert S. Cohen and Yehuda Elkana). 1977.
38. R. M. Martin, *Pragmatics, Truth, and Language.* 1979.
39. R. S. Cohen, P. K. Feyerabend, and M. W. Wartofsky (eds.), *Essays in Memory of Imre Lakatos.* 1976.
40. B. M. Kedrov and V. Sadovsky. *Current Soviet Studies in the Philosophy of Science.* Forthcoming.
41. M. Raphael, *Theorie des Geistigen Schaffens auf Marxistischer Grundlage.* Forthcoming.
42. Humberto R. Maturana and Francisco J. Varela, *Autopoiesis and Cognition. The Realization of the Living.* 1980.
43. A. Kasher (ed.), *Language in Focus: Foundations, Methods and Systems. Essays Dedicated to Yehoshua Bar-Hillel.* 1976.
44. Trân Duc Thao, *Investigations into the Origin of Language and Consciousness.* (Translated by Daniel J. Herman and Robert L. Armstrong; edited by Carolyn

R. Fawcett and Robert S. Cohen). 1984.
45. A. Ishimoto (ed.), *Japanese Studies in the History and Philosophy of Science*.
46. Peter L. Kapitza, *Experiment, Theory, Practice*. 1980.
47. Maria L. Dalla Chiara (ed.), *Italian Studies in the Philosophy of Science*. 1980.
48. Marx W. Wartofsky, *Models: Representation and the Scientific Understanding*. 1979.
49. Trân Duc Thao, *Phenomenology and Dialectical Materialism*. 1985.
50. Yehuda Fried and Joseph Agassi, *Paranoia: A Study in Diagnosis*. 1976.
51. Kurt H. Wolff, *Surrender and Catch: Experience and Inquiry Today*. 1976.
52. Karel Kosik, *Dialectics of the Concrete*. 1976.
53. Nelson Goodman, *The Structure of Appearance*. (Third edition). 1977.
54. Herbert A. Simon, *Models of Discovery and Other Topics in the Methods of Science*. 1977.
55. Morris Lazerowitz, *The Language of Philosophy. Freud and Wittgenstein*. 1977.
56. Thomas Nickles (ed.), *Scientific Discovery, Logic, and Rationality*. 1980.
57. Joseph Margolis, *Persons and Minds. The Prospects of Nonreductive Materialism*. 1977.
58. G. Radnitzky and G. Andersson (eds.), *Progress and Rationality in Science*, 1978.
59. Gerard Radnitzky and Gunnar Andersson (eds.), *The Structure and Development of Science*. 1979.
60. Thomas Nickles (ed.), *Scientific Discovery: Case Studies*. 1980.
61. Maurice A. Finocchiaro, *Galileo and the Art of Reasoning*. 1980.
62. William A. Wallace, *Prelude to Galileo*. 1981.
63. Friedrich Rapp, *Analytical Philosophy of Technology*. 1981.
64. Robert S. Cohen and Marx W. Wartofsky (eds.), *Hegel and the Sciences*. 1984.
65. Joseph Agassi, *Science and Society*. 1981.
66. Ladislav Tondl, *Problems of Semantics*. 1981.
67. Joseph Agassi and Robert S. Cohen (eds.), *Scientific Philosophy Today*. 1982.
68. Wuadysuaw Krajewski (ed.), *Polish Essays in the Philosophy of the Natural Sciences*. 1982.
69. James H. Fetzer, *Scientific Knowledge*. 1981.
70. Stephen Grossberg, *Studies of Mind and Brain*. 1982.
71. Robert S. Cohen and Marx W. Wartofsky (eds.), *Epistemology, Methodology, and the Social Sciences*. 1983.
72. Karel Berka, *Measurement*. 1983.
73. G. L. Pandit, *The Structure and Growth of Scientific Knowledge*. 1983.
74. A. A. Zinov'ev, *Logical Physics*. 1983.
75. Gilles-Gaston Granger, *Formal Thought and the Sciences of Man*. 1983.
76. R. S. Cohen and L. Laudan (eds.), *Physics, Philosophy and Psychoanalysis*. 1983.
77. G. Böhme et al., *Finalization in Science*, ed. by W. Schäfer. 1983.
78. D. Shapere, *Reason and the Search for Knowledge*. 1983.
79. G. Andersson, *Rationality in Science and Politics*. 1984.
80. P. T. Durbin and F. Rapp, *Philosophy and Technology*. 1984.
81. M. Marković, *Dialectical Theory of Meaning*. 1984.

82. R. S. Cohen and M. W. Wartofsky, *Physical Sciences and History of Physics.* 1984.
83. E. Meyerson, *The Relativistic Deduction.* 1985.
84. R. S. Cohen and M. W. Wartofsky, *Methodology, Metaphysics and the History of Sciences.* 1984.
85. György Tamás, *The Logic of Categories.* 1985.
86. Sergio L. de C. Fernandes, *Foundations of Objective Knowledge.* 1985.
87. Robert S. Cohen and Thomas Schnelle (eds.), *Cognition and Fact.* 1985.
88. Gideon Freudenthal, *Atom and Individual in the Age of Newton.* 1985.
89. A. Donagan, A. N. Perovich, Jr., and M. V. Wedin (eds.), *Human Nature and Natural Knowledge.* 1985.
90. C. Mitcham and A. Huning (eds.), *Philosophy and Technology II.* 1986.
91. M. Grene and D. Nails (eds.), *Spinoza and the Sciences.* 1986.
92. S. P. Turner, *The Search for a Methodology of Social Science.* 1986.
93. I. C. Jarvie, *Thinking about Society: Theory and Practice.* 1986.
94. Edna Ullmann-Margalit (ed.), *The Kaleidoscope of Science.* 1986.
95. Edna Ullmann-Margalit (ed.), *The Prism of Science.* 1986.
96. G. Markus, *Language and Production.* 1986.
97. F. Amrine, F. J. Zucker, and H. Wheeler (eds.), *Goethe and the Sciences: A Reappraisal.* 1987.
98. Joseph C. Pitt and Marcella Pera (eds.), *Rational Changes in Science.* 1987.
99. O. Costa de Beauregard, *Time, the Physical Magnitude.* 1987.
100. Abner Shimony and Debra Nails (eds.), *Naturalistic Epistemology: A Symposium of Two Decades.* 1987.
101. Nathan Rotenstreich, *Time and Meaning in History.* 1987.
102. David B. Zilberman (ed.), *The Birth of Meaning in Hindu Thought.* 1987.
103. Thomas F. Glick (ed.), *The Comparative Reception of Relativity.* 1987.
104. Zellig Harris *et al.*, *The Form of Information in Science.* 1987
105. Frederick Burwick, *Approaches to Organic Form: Permutations in Science and Culture.* 1987.
106. M. Almási, *Philosophy of Appearances.* Forthcoming.
107. S. Hook, W. L. O'Neill, and R. O'Toole, *Philosophy, History and Social Action. Essays in Honor of Lewis Feuer.* 1988.
108. I. Hronszky, M. Fehér, and B. Dajka (eds.), *Scientific Knowledge Socialized. Selected Proceedings of the Fifth Joint International Conference on History and Philosophy of Science Organized by the IUHPS, Veszprém, 1984.* Forthcoming.
109. P. Tillers and E. D. Green (eds.), *Probability and Inference in the Law of Evidence. The Uses and Limits of Bayesianism.* 1988.
110. E. Ullmann-Margalit (ed.), *Science in Reflection. The Israel Colloquium: Studies in History, Philosophy, and Sociology of Science.* 1988.
111. K. Gavroglu, Y. Goudaroulis, and P. Nicolacopoulos (eds.), *Imre Lakatos and Theories of Scientific Change.* 1989.
112. Barry Glassner and Jonathan D. Moreno (eds.), *The Qualitative-Quantitative Distinction in the Social Sciences.* 1989.

WIDENER UNIVERSITY
WOLFGRAM
LIBRARY
CHESTER, PA.